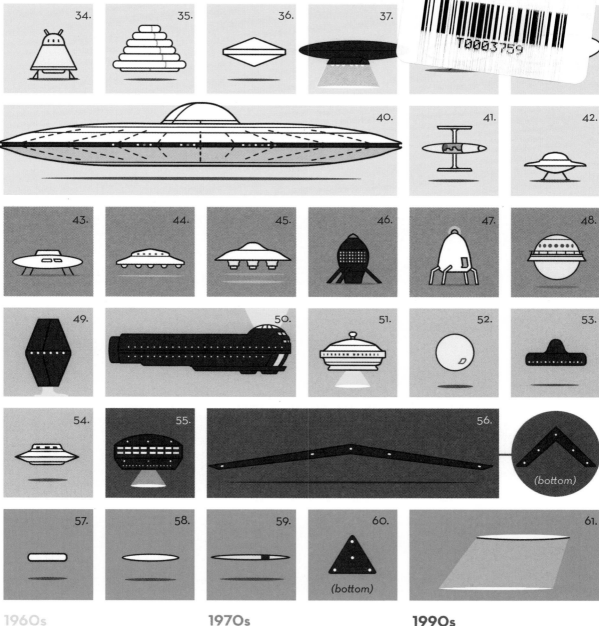

T0003759

1960s

31. Victoria, Canada, (1960)
32. Lancaster, USA, (1961)
33. Eagle River, USA, (1961)
34. Parr, UK, (1963)
35. Stanton, USA, (1964)
36. Merlin, USA, (1964)
37. Sunnyvale, USA, (1964)
38. Valensole, France, (1965)
39. Reboullion, France, (1966)
40. Vietnam, (1966)
41. Réunion Island, (1968)
42. Pirassununga, Brazil, (1969)

1970s

43. Kinnula, Finland, (1971)
44. Skipton, UK, (1978)
45. Charleston, USA, (1978)
46. Gerena, Spain, (1978)
47. Mindalore, South Africa, (1978)
48. Dechmont, UK, (1979)

1980s

49. Dayton, USA, (1980)
50. Mobile, USA, (1983)
51. Gulf Breeze, USA, (1987)
52. Voronezh, USSR, (1989)
53. Duclair, France, (1989)
54. Groom Lake, NV, USA (1989)

1990s

55. Yukon, Canada, (1996)
56. Phoenix, Arizona, (1997)

2000s

57. San Diego Coast, USA, (2004)
58. Chicago, USA, (2006)
59. French Coast, (2007)
60. West Midlands, UK, (2007)
61. Hangzhou, China, (2010)

To those I have annoyed with endless chatter about UFOs,
for the folk at Nobrow for joining in on the fun,
to David Clarke for his expertise,
and lastly, to the Greys from Zeta Reticuli; fly safely.

– Adam Allsuch Boardman

First published in 2020 by Nobrow,
an imprint of Nobrow Ltd. 27 Westgate Street, London E8 3RL.

Text and illustrations © Adam Allsuch Boardman 2020

Consultant: Dr David Clarke

10 9 8 7 6 5 4 3

Published in the US by Nobrow (US) Inc.
Printed in Poland on FSC® certified paper.

Also in the series:

MIX
Paper from
responsible sources
FSC® C163799

ISBN: 978-1-910620-69-4

www.nobrow.net

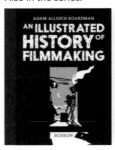

ADAM ALLSUCH BOARDMAN

AN ILLUSTRATED HISTORY OF UFOS

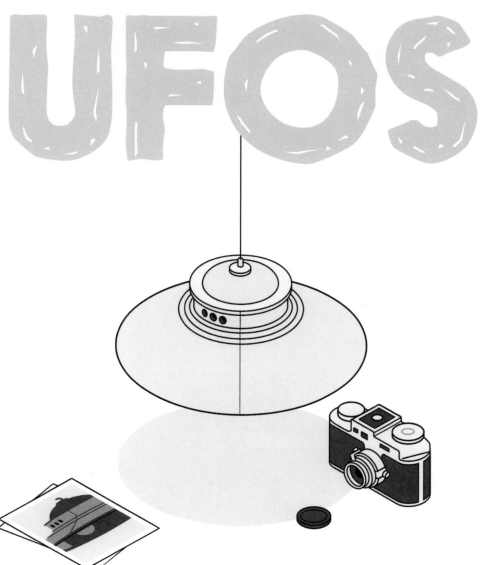

NOBROW

CONTENTS

Flanders, Belgium, 1972

INTRODUCTION

Throughout history, people have witnessed the spectacle of strange objects in the sky. Whether they are alien interlopers or mundane weather phenomena, their profound strangeness has spawned government enquiries and secretive societies, as well as inspiring countless dedicated investigators.

In the 1950s, The United States Airforce hoped to quantify the mystery with a clinical but influential title: 'Unidentified Flying Objects' (UFOs). Those who study the phenomena, therefore, are known as 'ufologists'.

When most think of 'UFOs', it conjures visions of flying saucers piloted by green aliens. This is but one explanation amongst many colourful and hotly debated hypotheses in which the aliens are rarely green.

This book has collected a wide range of close encounters, research and conspiracy theories, with the aim to inspire any intrepid UFO enthusiast to undertake further reading, documentary-viewing and patient sky-watching.

I have long held an interest in the phenomena thanks to the esoteric literature that haunted the shelves of book shops I frequented as a child. To properly research this particular book, I enjoyed a welcome return to the exciting and wild world of conspiratorial cork-boards, book collection and slightly frightening websites.

Ufology is an arena of fun speculation and debate, and my greatest hope is that you may go on to find your own truth, which is, after all, *out there*.

Common UFO shapes

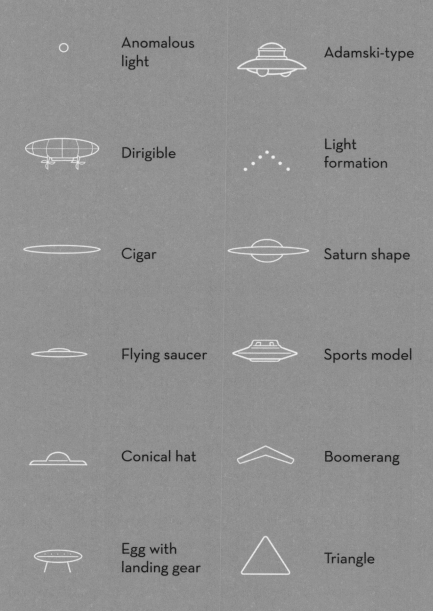

Anomalous light

Adamski-type

Dirigible

Light formation

Cigar

Saturn shape

Flying saucer

Sports model

Conical hat

Boomerang

Egg with landing gear

Triangle

Hitachi Province, Japan, 22 February 1803

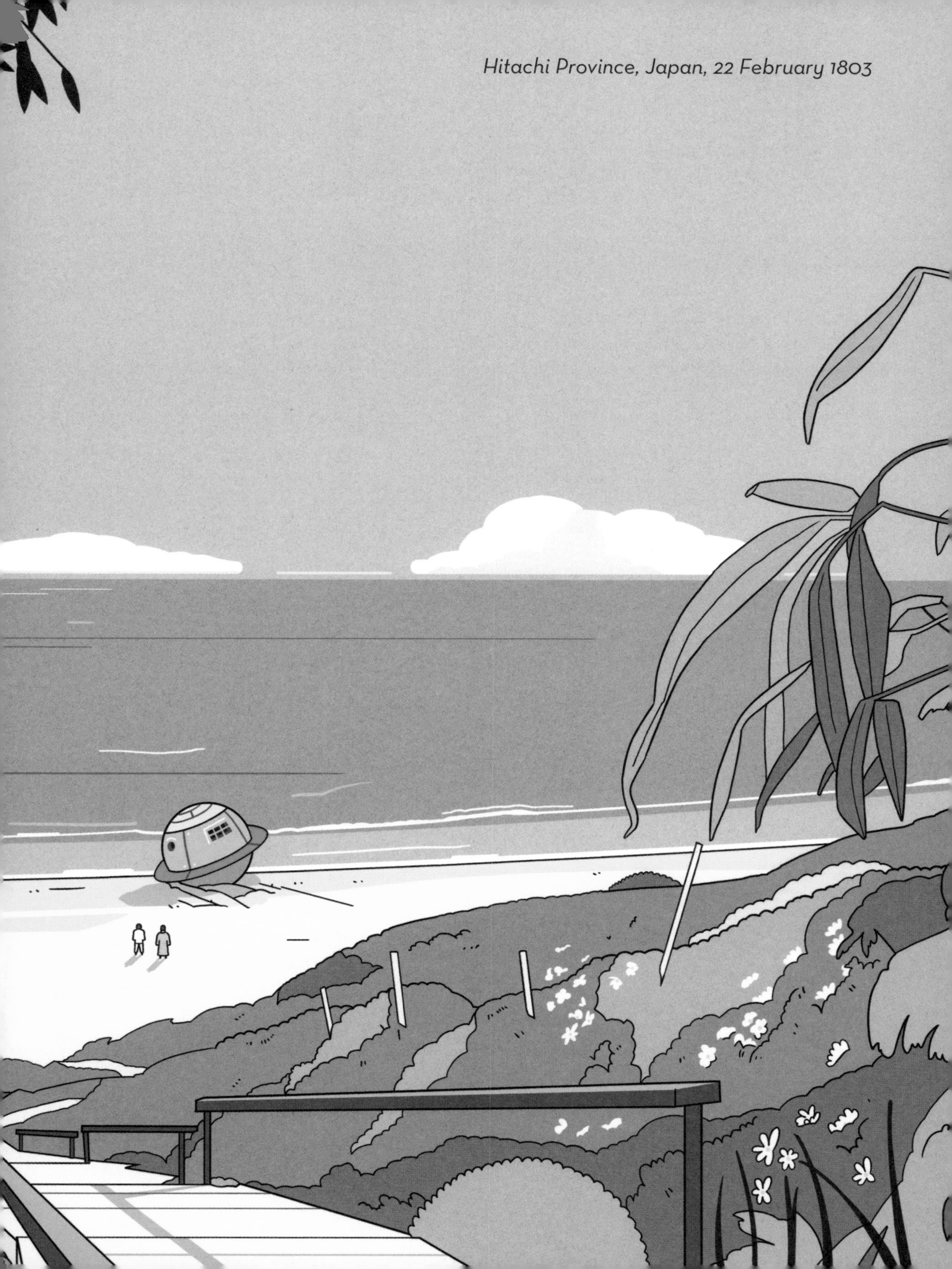

ANCIENT ASTRONAUTS

Humanity has always maintained an interest in the wider cosmos. Our ancestors built megaliths to align with stellar geometry and recorded the heavens by etching images onto stone and bone, and through petroglyph painting. Ancient explorers like the Polynesians and Vikings used the stars as a map to guide them to distant lands.

But what if the cosmos returned the interest? Some believe aliens visited ancient humanity, providing them with useful advice such as how to build cities and sow crops. The theory was proposed by the book *Chariots of the Gods?* (1968) written by Swiss hotel manager Erich von Däniken. The book alleges that mega-structures such as the Egyptian pyramids or the vast Nazca lines in Peru are evidence of extraterrestrial assistance, and of open dialogue between humans and aliens (the latter posing as gods).

Critics of the hypothesis suggest that it encourages a condescending, and often racist, attitude toward civilisations that earned their achievements through millennia of progress.

Erroneous evidence

The items on this page purportedly depict aliens, or show engineering knowledge beyond the reach of our ancestors.

Proponents of the ancient astronaut hypothesis believe that these kinds of artefacts are evidence of alien intelligence. As intriguing as this might be, such ideas tend to be dismissive of rich cultural histories and impressive craftsmanship.

Key

1. The Pyramids, Egypt
2. Stonehenge, UK
3. Shakoki-dogu figurine, Japan
4. Hieroglyphs, Egypt
5. Cave painting, Algeria
6. Petroglyph, USA
7. Moai, Easter Island
8. Antikythera mechanism, Greece
9. Aztek engraving, Mexico
10. Iron pillar, India
11. Quimbaya artefact, Colombia
12. Cave drawing, Italy
13. Painting, Serbia
14. Incan walls, Peru
15. The Nazca lines, Peru

1.

2.

3.

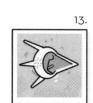
4.

5.

6.

7.

8.

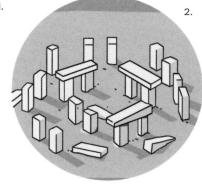
9.

10.

11.

12.

13.

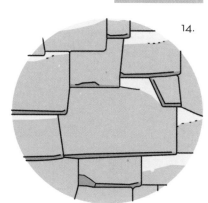
14.

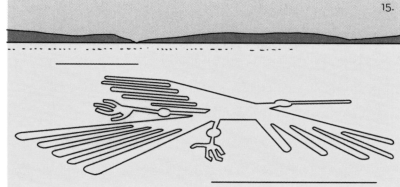
15.

MYTHIC VISITORS

Most cultures have their fair share of tales warning of færies, goblins and trickster spirits. While many consider the stories to be folk tales or myths, some believe they are accounts of actual alien encounters.

Middle Eastern god-aliens
In 1976, Russian author Zecharia Sitchin proposed that the Anunnaki, ancient Mesopotamian deities, were alien beings. In his books, Sitchin told of how the Anunnaki enslaved the human race for the purposes of extracting gold. These god-aliens returned to the planet Nibiru during the less agreeable ice age and this is where they presumably continue to plot against us.

Færie

European folklore frequently makes reference to mysterious other realms such as the land of 'færie'. Creatures from this realm, also called færies, are said to enter our world and kidnap unsuspecting humans. Some ufologists believe these stories stem from historic alien abductions.

Green children

In 12th century England, two odd siblings appeared in the village of Woolpit and confounded locals with their green hue. The younger child, a boy, died of sickness but the girl lived in the village for many years. Some suggest that the children were aliens left on Earth by some tragic circumstance.

Will-o'-the-Wisp

Will-o'-the-Wisps (meaning a torch that moves of its own volition) appear in folk stories across many cultures, from the Americas to East Asia. They are small, ghostly lights with the mischievous intent of drawing travellers from the road. Skeptics suggest the wisps are visions conjured by swamp gas or ergot poisoning from dodgy bread.

Cities in the sky

In the 19th century, locals in Orkney, UK and Finland reportedly saw cities floating in the firmament. US prospector Richard Willoughby claimed to have a photo of one such city above Alaska in 1899, but the picture was later found to depict a foggy day in Bristol, UK.

SPIRITS FROM SPACE

During the 19th century, spiritualism became incredibly popular in the West. The movement centred around the belief that the spirit world could be contacted through parlour activities (party games) or a 'medium' (a psychic individual).

Mediums claimed contact not only with earthly spirits, but also entities from other planets. During the 19th century, this usually meant Mars or Venus, as knowledge of more distant worlds was limited.

Medium to the Martians

Well-known American medium Vesta La Viesta (a pseudonym) gave many lectures in the early 1900s on her experiences of communing with Martians and Venusians through astral projection (the ability to move one's soul independent of the body).

Fun for the whole family

'Talking boards' such as the Ouija board were mass-produced from the 1890s, allowing any spiritual enthusiast to participate. Mass production may signify that the public was once more receptive to paranormal concepts, allowing for the introduction of ideas we may now baulk at.

Crowley's contact

British 'wizard' Aleister Crowley claimed to have made telepathic contacts during magic rituals. One such contact, in 1917, was with an entity named 'Lam'. Luckily, Crowley later illustrated this entity, which looks strikingly like a 'grey' alien (this was some decades before it became a popular image).

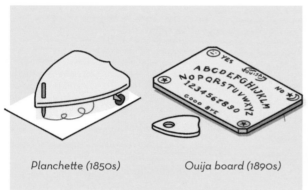

Planchette (1850s)   Ouija board (1890s)

Staging the supernatural

Proponents viewed spiritualism as a new science, while sceptics such as magician Harry Houdini exposed many mediums as frauds who exploited the grief-stricken for money and publicity. Techniques used by mediums included concealed strings to create the effect of levitation, extending grabbers for 'ghostly touches' and hidden music boxes for unearthly sounds.

MYSTERY AIRSHIPS

Towards the end of 1896, mysterious airships were sighted around California, USA. Mass sightings continued the next year across North America. These flaps (a period of many sightings) made excellent newspaper material. After compiling witness statements, some reporters suggested the airships may have arrived from Mars or Venus. The excitement over these airships may show how the new genre of science fiction influenced sightings.

Science fiction writers like Jules Verne and H.G. Wells wrote tales of airships and submarines, highlighting the public's fear of encroaching industrial progress. While more primitive balloons and underwater vehicles had been around for centuries, the 20th century saw these inventions manifest as symbols of terror, as airships led bombing raids and submarines struck at fleets.

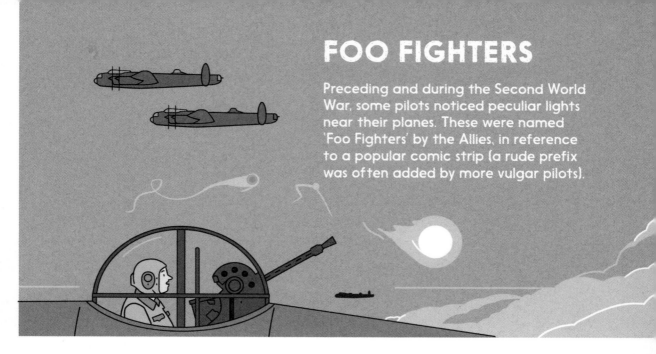

FOO FIGHTERS

Preceding and during the Second World War, some pilots noticed peculiar lights near their planes. These were named 'Foo Fighters' by the Allies, in reference to a popular comic strip (a rude prefix was often added by more vulgar pilots).

Wonder weapons?

The US 415th Night Fighter Squadron had several encounters with the Foo Fighters, and on one occasion chased a formation of orange fireballs, only to watch them eerily disappear as if turned off by a switch. Some suspected that Foo Fighters were secret weapons, a warranted concern when the war had introduced terrifying new technologies like Nazi V-2 rockets – the first long-range ballistic missiles.

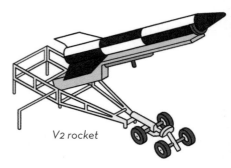

V2 rocket

Weird weather

After the war, St. Elmo's fire was commonly pointed to as the source of Foo Fighters. St. Elmo's fire is a weather phenomena that sometimes occurs in stormy conditions. When the air around pointy objects like wings or masts ionises, it can glow. The effect has been documented on modern aircraft, creating a web of harmless plasma around planes.

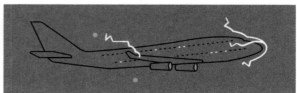

THE WAR OF THE WORLDS

On 30th October 1938, the Mercury Theatre in New York broadcast a play on CBS radio based on H.G. Wells's *The War of the Worlds* (1898). Directed and narrated by a young Orson Welles, the play was presented as a real news broadcast covering the invasion of Earth by Martians.

Mass hysteria

After reports that people believed there was an authentic alien invasion, reporters hounded the Mercury Theatre members and CBS staff, forcing Welles to issue a public apology. However, in all likelihood it gave his career a significant boost.

Pre-war pandemonium

Historians refute the scale of the panic and suggest CBS and the newspapers played up the chaos – it made fun news during a time when Europe was about to plunge itself into woeful war. Either way, it shows how eagerly the public engaged with the concept of extraterrestrial visitors.

WE SERVE DRINKS COOLED WITH
CARY GRANT KATHERINE HEPB
"B INGING UP BA

Fatal folly

A decade after the Mercury Theatre broadcast, an Ecuadorian adaptation by Radio Quito ended in tragedy. When the broadcasters revealed their 'news' had been fictitious, a mob of angry listeners arrived and set their building alight. The firefighters were late to the scene, having already been dispatched to deal with the faux Martians. Seven people tragically lost their lives in the flames.

THE 1940S

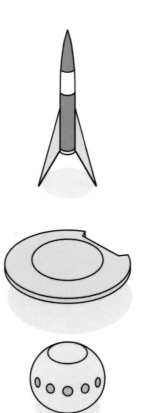

CRAFT TRAVEL 1200 MPH, SAYS FLIER
Shades of Flash Gordon!
'Saucers' Speed Over U.S.

Roswell Daily Record
RAAF Captures Flying Saucer
On Ranch in Roswell Region

SCIENCE FICTION FRENZY

The 1940s emerged as a golden age of science fiction (sci-fi), with pulp magazines, radio plays and film serials sparking the public imagination. Amongst the fiction, however, some authors penned allegedly true tales of cosmic mystery.

The Shaver Mystery

A progenitor of UFO mania began in 1943 when the eccentric Richard Shaver wrote to his favourite sci-fi magazine *Amazing Stories*. His letter detailed the alleged discovery of an ancient language called 'Mantong'. Impressed, the editor Raymond Palmer asked for more details. Shaver responded with a vast essay detailing his conviction that the ancient Teros people once lived in splendid cities underground. After significant editing and correspondence, Palmer published the story a few years later as a non-fiction piece in *Amazing Stories* called 'I Remember Lumuria!' (1945).

Secret Lumerian cave city

Deros creature

Devilish deros

Shaver's Lumerian mythos explained that the benevolent Teros evacuated Earth via rocket-ship, leaving the planet for humanity to inherit. He also sombrely warned that evil Deros creatures remained behind, lurking in the now-derelict cave cities. Deros made a sport out of torturing and abducting humans, either with mind control or more direct harassment

Shaver described how people were abducted by Deros from the subway or the sewers.

Early abductees

After the publication of Shaver's story, readers began writing into *Amazing Stories* claiming to have been abducted by Deros. Thrilled, Palmer published more Shaver-penned stories and anonymous testimonies.

The abduction stories are a clear precursor to the later alien abduction phenomenon. They both tend to involve technologically advanced beings tormenting human victims in a clinical or sexual manner. Eventually, Shaver-related tales decreased in popularity as UFOs claimed the spotlight in the late 1940s.

Victims of the Deros

Shaver's rock art

Later in life, Shaver became fixated with cutting rocks open and gazing at images in their interiors that only he could see. His paintings and photography based on the rocks garnered a degree of posthumous success. They are vivid illustrations of the Lumerian mythos, depicting Teros, Deros and lots of nudity.

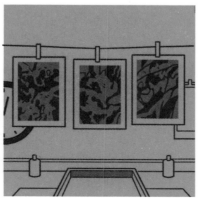

Shaver's rock art

FLYING SAUCERS

The modern history of UFOs is widely agreed to begin with the events of June 24th, 1947. On this particular day, Kenneth Arnold, a trader in fire safety equipment, observed something that would initiate the age of flying saucers. At around 3pm, Arnold was flying his CallAir plane near Mt. Rainer when he saw nine distant objects. Arnold vividly described the things as 'shaped like a pie-plate' and 'saucer-like' in their movement. When he reported his story to the press, they coined the enduring term 'flying saucer'.

A cultural icon

Sensing the potential for another sensational scoop, editor Raymond Palmer (of Shaver Mystery fame) contacted Arnold and asked to buy the story for his magazine, but Arnold happily gave an account for free. The issue's cover included an imaginative illustration of three flying saucers above Arnold's plane, establishing the flying saucer as a symbol of otherworldliness, synonymous with the UFO phenomenon.

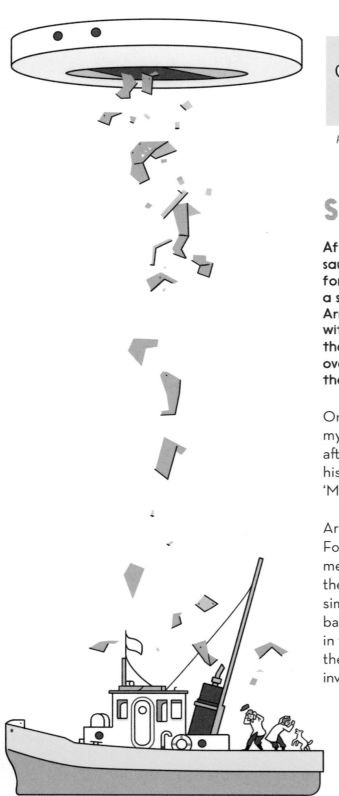

Flying Saucer

View from beneath

SAUCERS ATTACK!

After great public interest in Arnold's flying saucer sighting, Palmer sought similar stories for his new magazine, *Fate*. Learning of a similar sighting on Maury Island, he sent Arnold as a investigative journalist. He met with a pair of harbour workers who claimed that six doughnut-shaped craft had flown over their boat and dropped weird metal on them, killing a dog and injuring a young man.

One worker, Harold Dahl, claimed that a mysterious dark-suited man visited his trailer after the sighting and warned him not to share his tale, this being an early example of the 'Men in Black' legend.

Arnold excitedly called upon two US Air Force officers to take a look at the unusual metal left behind by the UFO. Upon arrival, the officers were quick to note that it was simply aluminium. In a hurry to get back to base for the next day, the officers took a flight in the early hours, but tragically died when the plane crashed, fuelling wild conspiracies involving a government cover-up.

THE ROSWELL INCIDENT

Shortly after Kenneth Arnold's flying saucer sighting in June 1947, a rancher named William 'Mac' Brazel found the debris from a crashed object near Roswell, New Mexico. Brazel thought the debris, mostly foil and rods, was curious. Once he reported his find in town, soldiers from the Roswell Army Air Field (RAAF) quickly arrived to spirit away the damaged detritus. The RAAF issued a press release on 8th July, announcing that they had indeed recovered a 'flying disc'!

Mac Brazel discovers debris

The RAAF announce they have recovered a 'flying disc'

Myths and legends

Details of the crashed spaceship, including dead aliens and the presence of the Men in Black passed into popular culture thanks to the seminal book *The Roswell Incident* (1980). The book was based on witness interviews gathered by authors Charles Berlitz and William Moore (helped by famous Canadian ufologist Stanton Friedman).

Project Mogul

The day after Brazel's discovery, the RAAF issued a correction: the recovered debris belonged to a weather balloon and not an alien spaceship. Historians now believe the reason for the rushed excuse was to conceal the testing of secret 'Project Mogul', a monitoring balloon used to spy on Russian atom bomb tests.

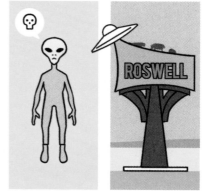

Saucer sightseeing

Whether it was a balloon or spaceship debris in 1947, Roswell has established itself as the spiritual capital of UFOs, with museums, themed restaurants, bars and highly inventive merchandise. Its name has become synonymous with themes of conspiracy, cover-up and aliens.

THE MEN IN BLACK

The Men in Black (MiB) are said to be the dedicated shadows of UFO witnesses. Although making their debut as far back in the 1940s, the myth was largely popularised by American author Gray Barker's book *They Knew too Much About Flying Saucers* (1953). Barker's book described strangers arriving at the homes of saucer enthusiasts to ask strange questions and make sinister threats.

The TV series Twin Peaks *takes a lot of inspiration from MiB stories, mixing UFO and occult lore.*

Funky phantoms

Witnesses have described the Men in Black as tall and unusually pale. Their faces are often gaunt and hairless, with lipstick applied to their thin, expressionless lips. To the general public, the MiB are most commonly associated with the *Men in Black* film franchise, in which they are depicted more like fun space cops than esoteric messengers of terror.

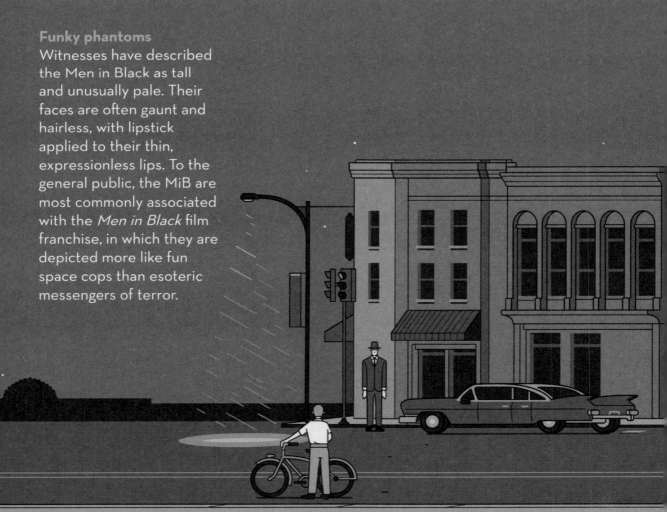

Sometimes they bring terror.
Orchard Beach, USA, 11th September 1976

You have two coins in your pocket. Take one out.

VANISH!

No one on this plane will see that coin again!

Destroy your UFO research, or suffer a similar fate!

Sometimes they're just weird.
Scarborough, UK, 1968

Can I see a glass of water?

What is this?

That was my dad's retirement present.

Is it your father's time? Is it here and now?

Your father! His time!

31

GOVERNMENT GETS INVOLVED

The US Airforce (USAF) began investigating UFOs in 1948. Their aims were to manage public interest, debunk cases and learn whether UFOs might be Soviet spy planes. Project Blue Book was their longest and best known project, coming after two divisive false starts.

Air Force investigations were often motivated by the fact their own personnel kept seeing UFOs around military bases.

Project Blue Book staff

1 Discover the truth

The first USAF project, **Project Sign** concluded in 1949 with a report called *Estimate of the Situation*, which outlined the remaining mystery of UFOs and entertained the possibility that some UFOs were spaceships piloted by aliens.

Holloman Airforce Base, USA

Alabama skies, USA, 24 July 1948

2 Deny everything

Dissatisfied with remaining questions, USAF leadership launched the second project, **Project Grudge** in 1949. Grudge sought to focus on debunking sightings and in August 1949, the Grudge Report stated that no UFO was beyond explanation and hoped that would end the affair.

❸ See both sides

Project Blue Book had a lasting impact on the UFO community as it introduced a lot of terminology and processes still used to this day. This investigation began in March 1952. Under the leadership of Captain Edward Ruppelt, the project hoped to provide more balanced assessment and apply a more scientific approach. Perhaps the most lasting impression has been made by the term 'UFO', coined during the project.

Cpt. Edward Ruppelt

> Obviously the term 'flying saucer' is misleading when applied to objects of every conceivable shape and performance. For this reason the military prefers the more general, if less colourful, name: unidentified flying objects, UFO (pronounced Yoo-foe) for short.

Capital craziness

After the significant flap of UFO reports in December 1952 centred around Washington DC, the CIA started a review of Project Blue Book. This 'Robertson Panel' formed in January 1953 and concluded the establishment must work to debunk rather than speculate, which unfortunately catalysed the impression amongst many civilian ufologists and journalists that the military was concealing the 'truth'.

Learned legacy

Astronomer J. Allen Hynek gained several decades of experience while consulting on USAF projects designed to explain UFOs. After the closure of Blue Book in 1969, he continued to be very active in ufology and set up the Centre for UFO Studies (CUFOS) in Illinois, USA.

THE BERMUDA TRIANGLE

In the study of the paranormal, many investigators have identified places in the world that seem to court strangeness. One famous example is the Bermuda Triangle, an area in the Atlantic Ocean where planes and ships seem to go missing in unusual ways.

Those who have survived their encounters with the Triangle report compasses going haywire, electrical fog, ghost ships and lights under the sea. Theorists have suggested many colourful explanations, such as time anomalies, technology from the lost city of Atlantis, and interference by flying saucers. Ufologists such as John Spencer and M.K. Jessup concluded that the Bermuda Triangle is plagued by malevolent aliens. They alleged that both planes and ships were drawn away from their correct paths by flying saucers, in a similar manner to færies misleading travellers in folklore.

Santa Maria, 1492

Shortly before making landfall in the Bahamas, the logs of the *Santa Maria* describe how the crew witnessed a ghostly light over the ocean waves, and later odd compass behaviour.

Carroll A. Deering, 1912

The schooner *Carroll A. Deering* was found abandoned just outside the Bermuda Triangle. The crew, lifeboats and provisions were missing but no clear reason could be discovered for the ship's abandonment.

Flight 19, 1949

In 1945, five US Navy planes, 'Flight 19', disappeared without a trace when training off the coast of Florida. Radio calls to base suggested an eerie interference; they had trouble navigating with compasses and couldn't spot familiar islands below. Radio contact was soon cut by poor weather, making their fate a mystery.

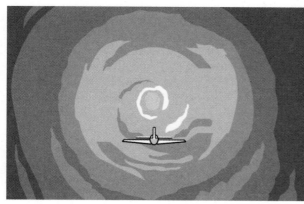

Time tunnel, 1976

While flying to Florida, Bruce Gernon and his father encountered a strange cloud. According to the Gernons, flying into the vapour revealed it to be a long tunnel filled with strange light. Upon exiting, they contacted air traffic control and found that they had arrived in half the usual time!

THE 1950S

Central Indiana, USA, October 1958

COMING OF THE CONTACTEES

The public enthusiasm for UFOs in the early 1950s catalysed a new variety of celebrity. Like spirit mediums before them, certain individuals stepped forth and proclaimed that they had communed with ambassadors from another world, except rather than the spirit world, these visitors came from space.

Alien encounters
In 1952, Polish-American, self-described spiritual teacher George Adamski claimed to have been visited by flying saucers and a charming alien named Orthon. Adamski detailed his encounters in his best-selling book *Flying Saucers Have Landed* (1953). The book was almost entirely written by Desmond Leslie, who later designed the first multi-track sound mixing desk. Adamski's tale was changed and elaborated upon through retellings, but is significant for popularising the character of the 'nordic' alien.

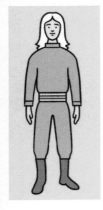

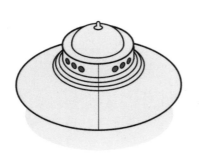

Alien archetype
Before the popular idea of aliens being short and grey, otherworldly visitors were typically reported to be light-haired folk in skin-tight suits. Initially they were referred to as 'Space Brothers', while today 'Nordic' is more commonly used. Orthon apparently made warnings about impending nuclear threat, but offered little practical advice.

Official intrigue
Project Blue Book's director Edward Ruppelt felt inclined to visit Adamski after receiving many letters requesting analysis of the account. Ruppelt found Adamski odd, and lamented in his UFO reports that increasing numbers of contactees were conning vulnerable people out of money.

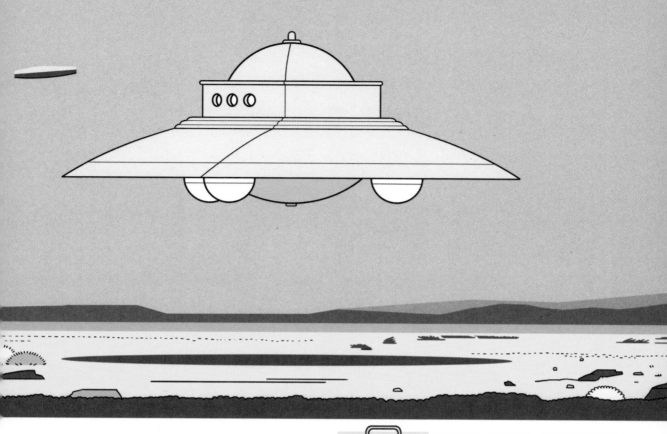

The Adamski Type

Adamski's book included blurry pictures of the alien craft, which detractors were certain exactly resembled part of a gasoline lamp. Regardless, the 'Adamski Type' flying saucer is an image that permeates popular culture, thanks in part to the fantastic cover illustration of Adamski's book.

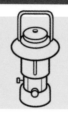

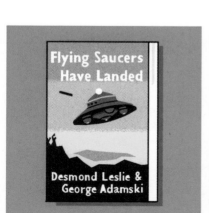

Scout Ship

Mother Ship

A change in the wind

The arrival of contactees like Adamski signified a tonal shift in UFO history. As many looked on and laughed at the apparent lunacy of those claiming to have met space brothers, ufologists found it increasingly difficult to be taken seriously.

SILVER SCREEN SAUCERS

After the public interest in cases such as Kenneth Arnold's sighting at Roswell, there followed copious science fiction films eager to cash in on the craze.

1.

2.

5.

3.

4.

6.

7.

8.

9.

10.

11.

Key

1. *The Flying Saucer* (1950)
2-3. *Flying Disc Man From Mars* (1950)
4-5. *The Day the Earth Stood Still* (1951)
6. *The Thing From Another World* (1951)
7-8. *Invaders From Mars* (1952)
9. *The War of the Worlds* (1953)
10-11. *Devil Woman From Mars* (1954)

12. *This Island Earth* (1955)
13. *Forbidden Planet* (1956)
14. *Warning From Space* (1956)
15-16. *Earth vs. the Flying Saucers* (1956)
17. *The Mysterians* (1957)
18-19. *Invasion of the Saucermen* (1957)
20-21. *Teenagers From Outer Space* (1957)
22-23. *Plan 9 From Outer Space* (1959)

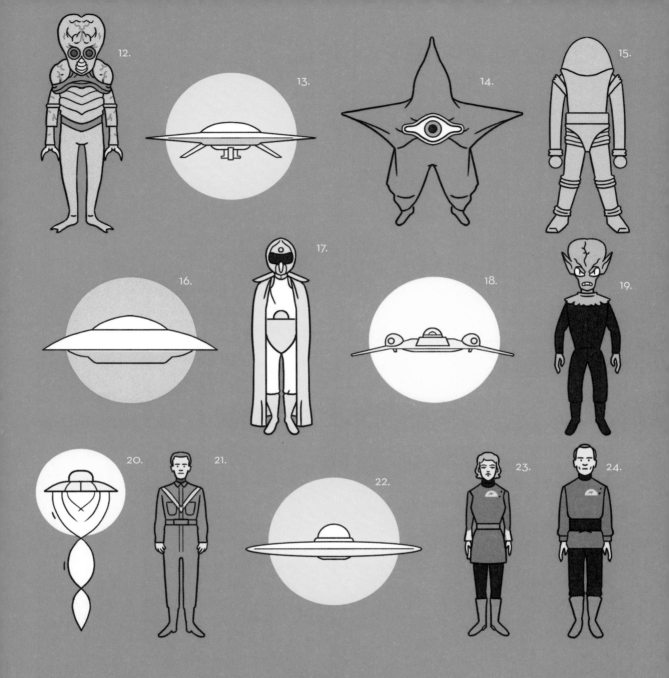

12. 13. 14. 15.

16. 17. 18. 19.

20. 21. 22. 23. 24.

Saucer semiotics

Films helped codify the image of the flying saucer into a universally recognisable design. The flying saucer prop, after all, can be made at low expense and rarely includes complex moving parts. It is possible that UFO witnesses, whether unconsciously or deceptively, coloured their recollections with what they had seen in familiar film depictions.

UFOS OVER THE PACIFIC

Not to be confined to America, UFOs began to be sighted all over the globe, prompting the agencies of respective nations to undertake their own investigations.

Rockhampton, Australia, 14th August 1952

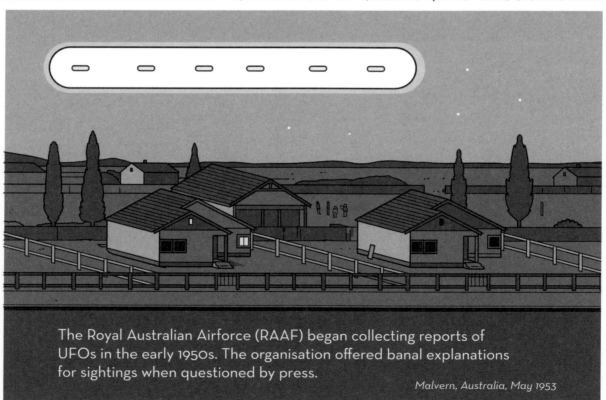

The Royal Australian Airforce (RAAF) began collecting reports of UFOs in the early 1950s. The organisation offered banal explanations for sightings when questioned by press.

Malvern, Australia, May 1953

The Drury Affair

One case in 1953 drew particular press attention due the footage captured by aviation official Tom Drury.

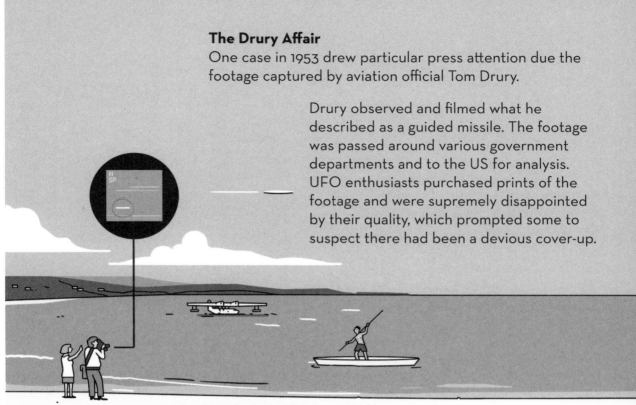

Drury observed and filmed what he described as a guided missile. The footage was passed around various government departments and to the US for analysis. UFO enthusiasts purchased prints of the footage and were supremely disappointed by their quality, which prompted some to suspect there had been a devious cover-up.

Port Moresby, Papua New Guinea, 23rd August 1953

Delightful disclosure

In 1982, ufologists were delighted to be granted access to the RAAF's UFO files. The files, now gathered and studied by many UFO historians, allow us to see the exchanges between officials and civilian groups. Responding to demands for an investigation of 'The Father Gill Sighting' (a missionary in Papua New Guinea saw aliens waving from a flying saucer) officials dryly suggested the event was 'probably... natural phenomena'.

To the disappointment of some believers, disclosures such as this are not the long-sought smoking guns, but simply expose a dull truth – militaries know little more about UFOs than the general public. This is perhaps why many favour more exciting conspiratorial narratives.

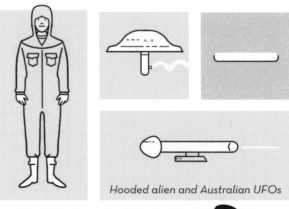

Hooded alien and Australian UFOs

BRITAIN'S FLYING SAUCER WORKING PARTY

In 1950, the UK's Ministry of Defence (MoD) created 'The Flying Saucer Working Party' to look into UFO sightings. The committee had five members, each representing various intelligence branches such as the Air Ministry and War Office.

The British perspective

The MoD decided that out of hundreds or reports from around the world, only three were worthy of a raised eyebrow, perhaps ascribing their value based on the witnesses being male military personnel. Regardless, they eventually dismissed each case as optical illusions.

Humbug!

After a few months the men produced a six-page report with the catchy title 'DSI/JTIC'. The report was doubtful of UFO cases, and cast suspicion on the mental faculties of witnesses. However, rejecting the existence of UFOs did not make people stop reporting them.

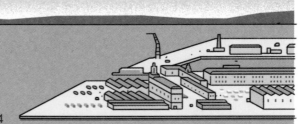

RAF Tangmere & Portsmouth, UK, 1ˢᵗ June 1950
UFO tracked by RADAR and and sighted by a jet pilot

15th August, 1950, Farnborough, UK
Saucer seen above air field

Bemused bosses

A well reported July 1952 flap of UFOs in Washington D.C. prompted a confused memo from Prime Minister Winston Churchill on 25th July 1952. This is surprisingly not unusual, many leaders of state have made official enquiries into UFOs, indeed, prior to becoming US president Jimmy Carter claimed to have seen one in 1969.

> What does all this stuff about flying saucers amount to? What can it mean? What is the truth?

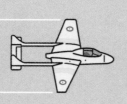

The UFO desk

After the working party ended, UFO reports were forwarded to the Air Ministry and later the Ministry of Defence, to investigate and to liaise with press and the public. Over time, this informal department became known as the 'UFO desk' and endured until the 1990s.

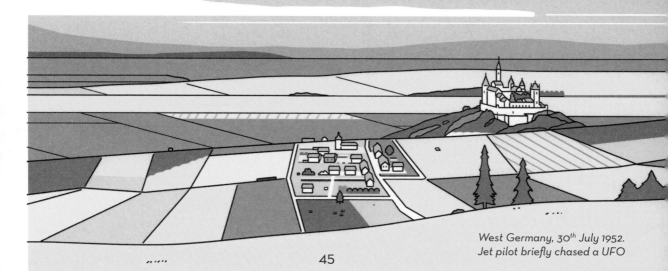

West Germany, 30th July 1952.
Jet pilot briefly chased a UFO

GIANT ROCK SPACECRAFT CONVENTION

As UFOs became an object of popular speculation and interest to the public (thanks largely to science fiction novels and films), a subculture emerged of those who believed in their extraterrestrial origin. From 1953, American aircraft mechanic and contactee George Van Tassel hosted an annual 'Giant Rock Spacecraft Convention', at his airfield in California.

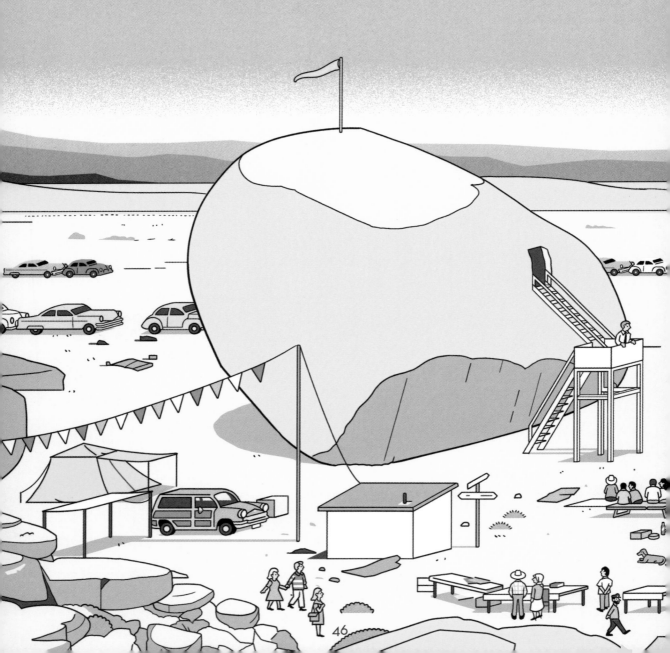

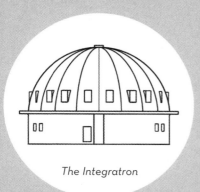

The Integratron

Community gathering

The convention was an important exchange of UFO ideology, largely in favour the extraterrestrial angle. Thousands of people attended up until the 1970s to hear lectures and trade the latest UFO books and paraphernalia. The convention attracted guests until Van Tassel's death in 1978, by which point many other conventions had established their own communities.

A home for meditation

The UFO community had many ties to new age and esoteric thought. Not far from the convention ground Van Tassel built a structure called 'The Integratron' (a vague sciencey name). The build was funded largely by donations, including some by American tycoon Howard Hughes. The structure was believed to aid in meditation and allegedly cure ailments, and remains open today.

SPACE INVADERS

During the 1950s there was a boom of sightings that involved a 'UFOnaut' or alien. One example on 12th September 1952 involved seven witnesses to a malevolent mechanical alien in Flatwoods, West Virginia, USA.

Upon seeing a fireball fall into the woodland, a party of locals rushed into the woods where, through thick fog, they saw a large creature with shining eyes. Terrified, they ran home and later gave accounts to the press. Aspiring ufologist Gray Barker was soon on the scene conducting interviews. Sceptics believe the group were merely startled by an owl.

THE ORIGINAL ABDUCTEE

On the warm night of 16th October 1957, the farmer Antônio Vilas Boas was ploughing crops in Minas Gerais, Brazil, when he apparently spied a flying saucer land nearby. According to Boas, three tall figures in spacesuits rushed out to grab him, and he was dragged into the saucer.

Cosmic kidnapping

Inside, Boas was subjected to unusual medical procedures that made him vomit. Following this poor treatment, an alien woman approached him and made sexual advances.

Boas, having been traumatically assaulted, was returned to his spot of abduction, finding that many hours had passed unaccounted for. After the incident, Boas got in touch with the journalist Jose Martins, after he spotted an advert in the paper asking for UFO experiences. Boas's story is representative of the many traumatic abduction experiences that emerged after Adamski reported on his own apparently amiable encounter.

Alien assailants

49

THE 1960S

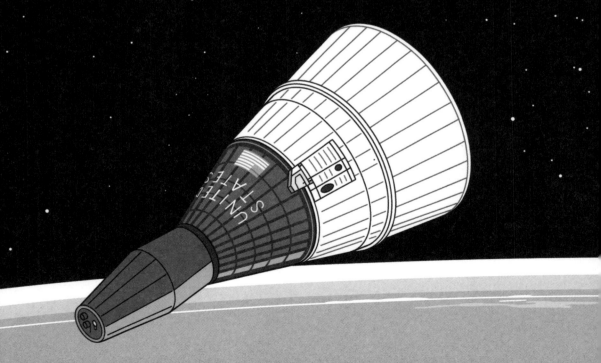

Earth's orbit, June 1965

THE HILL ABDUCTION

The 1960s were a veritable boom time for the phenomenon, signified by high profile abduction cases, mass sightings and increasingly unusual case details. All the while, official government enquiries did little to stem public interest or entirely explain away UFO activity.

An interrupted journey

The most well-known abductee incident concerns the American couple Betty and Barney Hill. Their experience, recovered through hypnosis, is credited with popularising the 'grey alien' archetype. On 19th September 1961, the Hills were driving back from a holiday when they spotted a flying saucer. Having studied the craft closely, they left with a sense of unease. When the Hills returned home they had the unusual impression that they had forgotten something important.

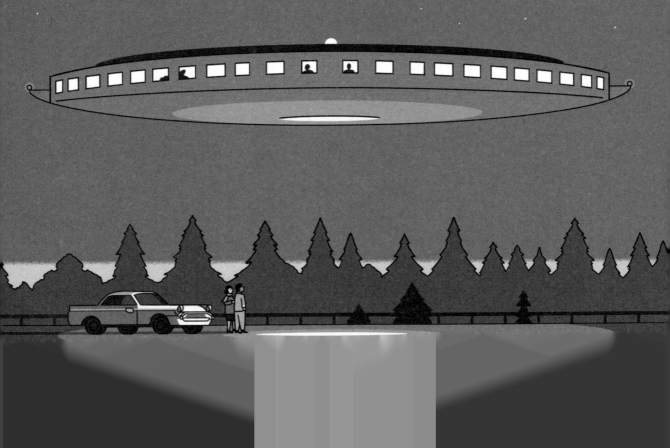

Making a report

Betty telephoned Pease Airforce Base to report the sighting. The officers decided the couple had misidentified Jupiter.

Betty does her research

Still eager to make sense of what the two of them had seen, Betty borrowed a UFO book from the library by Major Donald Keyhoe – director of the civilian UFO research group NICAP – and decided to write to him about her experience.

Donald Keyhoe, director of NICAP

Missing time

In response to Betty's letter, NICAP investigators visited the Hills and found them to be a delightful and honest couple. They noted that the Hill's drive home had taken much longer than it ought to have and estimated that the couple had misplaced *seven* hours after the encounter!

An element of 'missing time' has been noted across many abduction accounts. Some ufologists attribute it to a deliberate attempt by the alien abductors to repress memories of the abduction.

Under hypnosis

After the UFO sighting, Betty suffered from unsettling dreams, while Barney felt persistent physical aches. Hoping to address their maladies, the Hills visited hypnotist Dr. Benjamin Simon. During seven months of hypnotic sessions, the couple recalled the horrifying abduction experience that followed their UFO sighting.

Bestseller!

The Hills were able to separately recall similar accounts of being taken from the road by short, grey aliens. These aliens subjected the Hills to demeaning medical tests aboard their spaceship. The Hills' accounts were publicised in author John Fuller's bestselling book *An Interrupted Journey* (1966).

The spaceship, based on the Hills' rough sketches

Distant stars

The book included a sketch of a star map seen by Betty. Reader Marjorie Fish found it similar to the obscure star system of Zeta Reticuli. Detractors believe the similarities are coincidental. Regardless, Zeta Reticuli would be referenced by future abductees and in popular culture (planet LV-426 in the 1979 film *Alien* is located there).

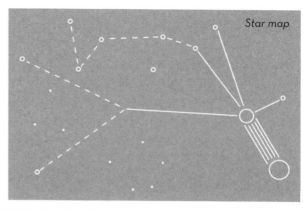

Star map

According the Hills the grey aliens wore sleek, militaristic uniforms.

The greys

The aliens described by the Hills took the form of the now-familiar grey alien: a short, pale creature with a large head and eyes. This appearance was not totally original, having being described by some witnesses before. However, the extremely well publicised Hill story established greys as the de facto popular conception of 'alien'.

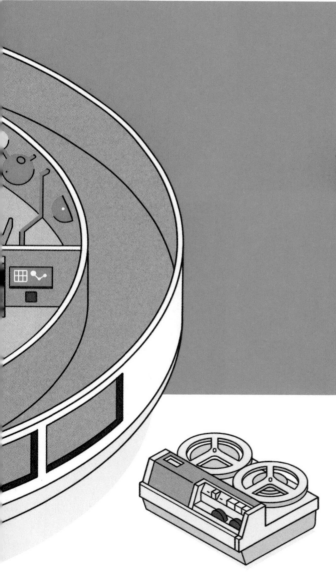

Recorder used for hypnosis sessions

Too much TV?

Skeptics such as Marty Kottmeyer have suggested the Hill's hypnotic recollections were nothing more than a kaleidoscopic retelling of media they had recently viewed. Most damningly, two weeks before their hypnosis an episode of science fiction series *The Outer Limits* had featured aliens that looked startlingly similar to the ones described by the couple.

IDENTIFIED FLYING OBJECTS

Throughout the 1960s, the US Air Force's Project Blue Book continued to debunk UFOs with increasingly creative explanations for the press, often going out of its way to avoid the extraterrestrial hypothesis. One infamous explanation was to blame sightings on 'swamp gas', which became an in-joke within ufology.

LENTICULAR CLOUD

REFLECTIVE BIRDS

SWAMP GAS

PLANET

SHOOTING STAR

AURORA BOREALIS

BALL LIGHTNING

LIGHTNING SPRITE
(Upper atmosphere lighting)

ST. ELMO'S FIRE
(See: Foo Fighters p.19)

Ambiguous aircraft

Meanwhile, the Cold War between America and the USSR quietly raged on. An arms race of hi-tech aircraft escalated, and modern ufologists believe that these were responsible for a great many sightings.

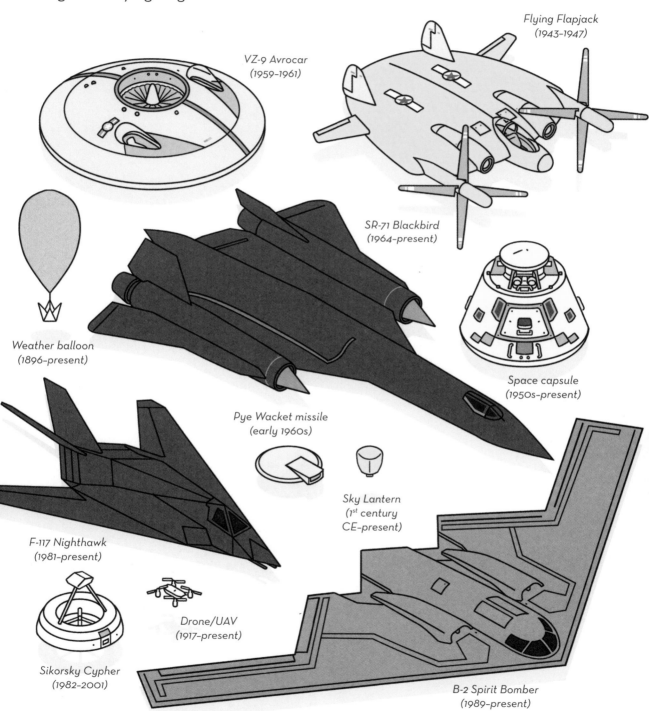

Flying Flapjack
(1943–1947)

VZ-9 Avrocar
(1959–1961)

SR-71 Blackbird
(1964–present)

Weather balloon
(1896–present)

Space capsule
(1950s–present)

Pye Wacket missile
(early 1960s)

Sky Lantern
(1st century
CE–present)

F-117 Nighthawk
(1981–present)

Drone/UAV
(1917–present)

Sikorsky Cypher
(1982–2001)

B-2 Spirit Bomber
(1989–present)

CIVILIAN UFO ASSOCIATIONS

Outside of official government and military interest in the matter, many civilian and amateur groups were forming as public interest in UFOs swelled in the 1950s and 60s.

Members of associations in this era were not entirely composed of fringe theorists or even people that believed that UFOs were aliens. They were largely just curious people with an interest in learning more about the mystery.

UFO associations were usually reliant on donations and subscriptions to their journals (and thus were often short on money). UFO journals reported on contemporary sightings, and printed the latest photos as well as reader letters.

NICAP

The American 'National Investigations Committee On Aerial Phenomena' (NICAP) formed in 1954. It attracted a membership of moderate UFO enthusiasts – those who were agnostic on UFOs being alien. To avoid ridicule from the broader public, NICAP investigators usually refrained from looking into the more esoteric areas of ufology, such as abductees and alien sightings. NICAP suffered from money troubles until dissolving in the 1980s.

BUFORA

In 1962, the 'British UFO Research Association' (BUFORA) formed to coordinate the efforts of the many UFO groups (known as 'Saucer Clubs') around the UK. BUFORA member John Spencer wrote the exhaustive *UFO Encyclopaedia* (1991), a guide to the most provoking UFO cases and individuals. BUFORA remains active to this day.

MUFON

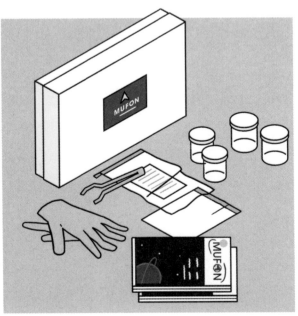

The American group 'Mutual UFO Network' (MUFON) was formed in 1969. Many of their members were united by a dissatisfaction with the government's begrudging attitude towards the phenomenon. MUFON released periodic field manuals from the early 1970s, to ensure a degree of consistency in field investigators. Nowadays, they offer tiered paid membership, enabling exclusive access to their archives. MUFON remains one of the largest civilian UFO investigation organisations.

ACROSS THE DIMENSIONS

Up until the 1960s, theories regarding
UFOs were split between mundane or alien
hypotheses. However, in the 1960s, authors such
as John A. Keel and French astronomer Jacques
Vallée proposed the ideas that UFOs, and the
paranormal more broadly, were a symptom of
things crossing over from different realities.
This is called the 'interdimensional hypothesis'.

Jacques Vallée

1 Interdimensional event
The process begins with a witness
to an interdimensional event
beyond human comprehension
(think a David Lynch film). The
event could be natural or
intentionally caused by intelligent
beings from another reality.

2 Filling in the gaps
To process this confusing event,
the witness explains it to
themselves using the social and
cultural symbols of their time.
This could be an angel, ghost,
devil, alien or UFO.

3 Sharing the story
Finally, the witness tells their story
to their friends, family or the media.
This bolsters the prominence of
cultural symbols; when people
claim to see UFOs it encourages
others to explain their paranormal
experiences with similar imagery.

MILITARY OPS

Psychological operations (PSYOPs) are military actions intended to influence opinions, ranging from propaganda to the cruel use of music to torture prisoners. Ufologists such as Vallée suggested that some agencies stage supernatural phenomena (including UFOs) to misdirect and terrify targets.

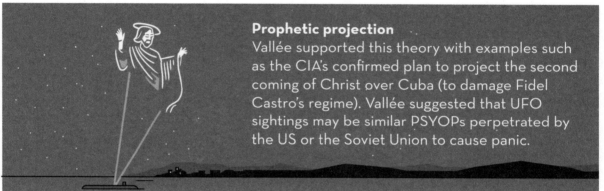

Prophetic projection
Vallée supported this theory with examples such as the CIA's confirmed plan to project the second coming of Christ over Cuba (to damage Fidel Castro's regime). Vallée suggested that UFO sightings may be similar PSYOPs perpetrated by the US or the Soviet Union to cause panic.

MKULTRA

In the Cold War, the CIA's MKUltra and the US military's Edgewood programs used psychedelic drugs and sensory torture in an attempt to produce super-soldiers and mind-control weapons. Conspiracy-keen ufologists suggest some UFO experiences were induced by experiments on victims. The series *Stranger Things* (2016–present) loosely explores this concept.

Alleged Soviet saucer

Ballistic missile

Soviet saucers
Some theorists believe UFOs were Soviet Union hoaxes to make Americans paranoid. The Soviets did accidentally cause UFO sightings at home when testing secret R-36 ballistic missiles in 1967. During 1950s and 1960s the FBI investigated individuals in the UFO community, but found no evidence of 'communist subversion'.

THE KECKSBURG INCIDENT

On the evening of 9th December 1965, a strange object allegedly smashed into the Kecksburg woodland in Pennsylvania, US. The press reported that a thorough search had found no trace of any fallen object.

On 19th September 1990, an episode of television series *Unsolved Mysteries* rekindled interest in the story, and alleged that the fallen object had actually been an acorn-shaped spacecraft that was quickly smuggled away by the US military.

NASA findings

Searching for the truth in the 2000s, journalist Leslie Kean forced NASA to share hundreds of files by means of a lawsuit sponsored by the Sci Fi channel. Kean's hope was that the files would unearth tangible proof that a space object had been taken from Kecksburg for secretive study. Disappointingly, the files did not unveil a shadowy conspiracy, but did contain some mention of NASA providing

Experimental craft?

Re-entry capsule?

Satellite or super-weapon?

Skeptics generally agree that witnesses to the Kecksburg event probably saw part of a fallen satellite or meteor being whisked away by the USAF. In conspiracy corner meanwhile, The History Channel's imaginatively named documentary 'Nazi UFO Conspiracy' proposed that the acorn-shaped object was actually a flying saucer built by Nazis. The story goes that the USAF secured the craft at the end of WW2 but can't get it to fly without periodically crashing into the woods.

THE ANDREASSON ABDUCTION

Following the well-reported abduction of Betty and Barney Hill, the 1960s were busy with abductee claimants. On 25th January 1967, Betty Andreasson experienced a particularly wild visitation by aliens.

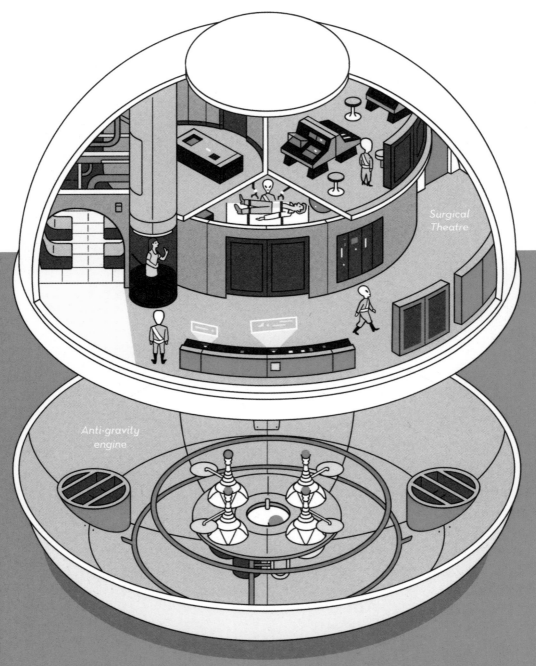

Surgical Theatre

Anti-gravity engine

Religious aliens

Andreasson's case began one evening when grey aliens dramatically floated through the kitchen wall and froze her family with alien wizardry. Andreasson was treated to a trip on the flying saucer and discovered the aliens worshipped Jesus Christ. After medical examinations and transcendent spiritual experiences, she was kindly returned home.

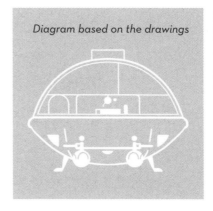

Diagram based on the drawings

A scene from the drawings

Abductee art

While many abductees have attempted to render their experiences with sketches or sculpture, few have achieved imagery quite as vivid as the illustrations provided by Andreasson. She applied her talents to describe her alien experiences as well as the exact layout of a flying saucer.

High strangeness

Within the study of UFOs, ufologists apply to term 'high strangeness' to cases such as Andreasson's that exhibit a surreal, dream-like quality. Some claim this is evidence of witness sincerity (why make it harder for oneself to be believed)? Others suggest it is the contrary: evidence of improvised falsehood.

A result of trauma?

Sceptics such as Dr Aaron Skulich delicately noted that Andreasson's emergence as an abductee coincided with a personal loss. He believes the religious aspect of her experience is telling, suggesting that it may have been a symptom of grief.

Diverse aliens seen by Andreasson

THE 1970S

Pascagoula, USA, 20th October 1973

UFO CULTS

The mid-century was a time of social revolution and the widespread use of recreational drugs. This led to the formation of new cults and religions, some of which involved an interest in UFOs and alien prophets.

A few high-profile examples were associated with murder or acts of mass-suicide, while the more innocuous expressed themselves through fashion and ritual, sticking closer to mainstream religious practice.

IAM
est. 1930, USA

George Adamski
est. 1930s, USA

George Van Tassel
est. 1950s, USA

Unarius Academy of Science
est. 1954, USA

FIGU
est. 1970s, Switzerland

Raelism
est. 1974, France

Heaven's Gate
est. 1974, USA

One World Family
est. 1970s, USA

Chen Tao
est. 1990s, China

Jakarta's Eden
est. 2000s, Indonesia

Celestial chic
The Unarius Academy of Science was oneof many UFO religious groups to emerge from California. They believe in reincarnation (their late leader Ruth Norman apparently had over 200 lives) and await the arrival of an alien space fleet. They are still active and are best known for incredibly imaginative fashion and videos.

Heaven's Gate tragedy
In March 1997, 39 members of the American UFO cult Heaven's Gate took their own lives, at the insistence of leader Marshall Applewhite. They believed the passing Hale-Bopp comet was a sign of imminent alien arrival, and that their deaths would lead to a rapture-like transcendence. Their website is maintained by surviving members to this day.

FREAKY FORESTS

Forests are a common setting for folklore, and ever since the Flatwoods Monster there have been many sightings of UFOs and their alien occupants trespassing in dense woodland. Some witnesses have described aliens taking samples of flora and leaving hurriedly after being discovered.

At loggerheads
In 1971, Finnish loggers Petter Aliranta and Esko Sneck were finishing for the day when they were interrupted by a landing saucer. Startled by the emergence of a green-suited creature, the loggers charged with their chainsaws, prompting the alien to blast off hurriedly. We should be thankful the aliens did not return the greeting by setting their own weaponry on Finland.

CLOSE ENCOUNTERS

Project Blue Book's Professor J. Allen Hynek published his book *The UFO Experience* in 1972, detailing his methodology for classifying the quagmire of UFO reports. The Close Encounters system, more commonly referred to among ufologists as the 'Hynek scale', is still used today.

J. Allen Hynek

1
Night lights

The most common and least strange UFOs: lights in the night sky.

2
Daylight discs

UFOs seen in the daytime. These are most commonly discs or saucers.

3
Radar-Visual

A visual encounter corroborated with RADAR contact.

Heroic Hynek

Hynek's contributions to ufology are an inspiration to others in the field. While some lambasted him for providing unimaginative explanations for UFO sightings, his foremost interest was being scientific. He was less eager to push alien conspiracy theories and instead exhibited delightful enthusiasm for the puzzle that UFOs presented.

Cinematic consultant

Hynek was consulted for Steven Spielberg's film *Close Encounters of the Third Kind* (1977). He even had a cameo during the film's climax, in which he is seen in a crowd of on-lookers with his characteristic pipe. The Ufologist character played by director François Truffaut is said to be based on either investigator Claude Poher or Jacques Vallée.

4

Close Encounter of the First Kind

The UFO is seen in close proximity.

5

Close Encounter of the Second Kind

The UFO has a physical effect, such as leaving marks or stopping car engines.

6

Close Encounter of the Third Kind

UFO occupants are seen in or around the craft.

THE TRAVIS WALTON INCIDENT

On the evening of 5th November 1975, lumber worker Travis Walton and six of his colleagues were driving through the forests of Snowflake, Arizona, USA. Upon seeing a UFO, the group pulled over to take a better look.

Walton stepped out of the truck and was apparently struck by a strange light, prompting his friends to pull away in blind panic. This marked the beginning of his apparent five-day disappearance.

After fruitless searching by police and volunteers, Walton announced his 'return' in a call from a remote phone booth. The police believed the affair to be a hoax.

Nordics and greys

Upon his return, Walton explained that an alien abduction accounted for his disappearance. He recalled details of menacing greys, and his wanderings around the spaceship until he was returned to Earth by some benevolent Nordic aliens.

Polygraph test

Walton was scrutinised by polygraph tests at the behest of the Aerial Phenomena Research Organization (APRO). While Walton failed an initial test, he felt the technician had been deliberately rude to him, thus throwing the results. Walton has since passed other tests, although polygraphs are no longer widely recognised as indicators of truth.

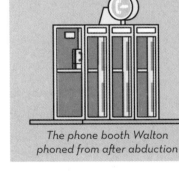

The phone booth Walton phoned from after abduction

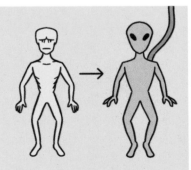

Aliens from Fire in the Sky *wearing suits that give them the iconic grey look.*

Legacy in the media

Walton's story was scintillating material for the press and made an impression on the broader public due to the lengthy disappearance and witnesses to his abduction. Walton wrote an account of his alleged abduction in *The Walton Experience* (1978), which was loosely adapted into the chilling horror film *Fire in the Sky* (1993).

IS BIGFOOT AN ALIEN?

The ape-like Bigfoot creature is an example of a cryptid, an animal whose existence is questionable. Those who specialise in the study of such mythic creatures are known as 'cryptozoologists'. Bigfoot has been the subject of sightings since the 19th century, while similar ape-like creatures have featured in folklore all over the planet.

Bigfoot crossed into the realm of UFOs in 1973, when witness Reafa Heitfield saw the creature enter a flying saucer outside her trailer home in Cincinnati, USA. Many cryptozoologists are partial to the belief that cryptids are cosmic in origin, either abandoned on Earth or the remnants of weird alien experiments.

A quick cryptid guide

1. Hairy dwarf, Venezuela
2. Skunk ape, USA
3. Bigfoot, USA
4. Mongolian death worm, Gobi Desert
5. Batsquatch, USA
6. Pope Lick monster, USA
7. Yeti, Himalayan mountains
8. Ahool, Java
9. Crawfordville monster, USA
10. Gef the talking mongoose, UK
11. Dover demon, USA
12. Chupacabra, USA
13. Dogman, USA
14. Loveland frog, USA
15. Cabbagetown tunnel monster, USA
16. Jersey devil, USA
17. Spring-Heeled Jack, UK

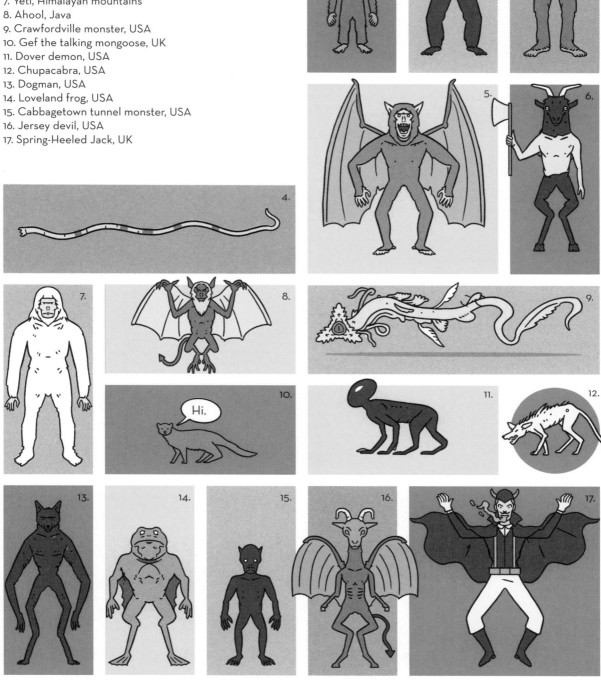

THE PETROZAVODSK INCIDENT

In the early hours of 20th September 1977, those around Petrozavodsk, USSR were treated to a brilliant UFO light display. The object took the form of a luminescent jellyfish that rained forth beams of light.

Cold War worries
According to journalists such as Nikolai Milov, some witnesses threw themselves under cover in terror, believing the display was the result of the long-dreaded nuclear apocalypse.

Soaring satellite?

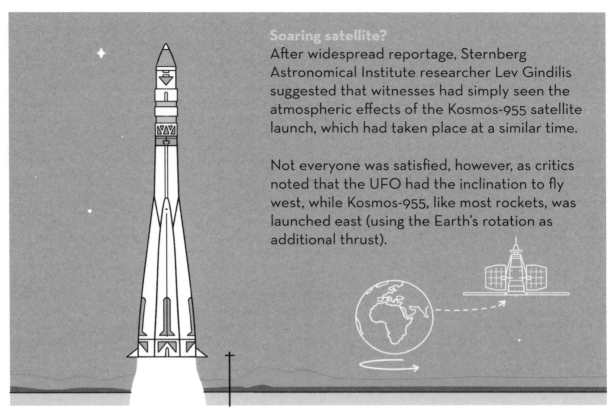

After widespread reportage, Sternberg Astronomical Institute researcher Lev Gindilis suggested that witnesses had simply seen the atmospheric effects of the Kosmos-955 satellite launch, which had taken place at a similar time.

Not everyone was satisfied, however, as critics noted that the UFO had the inclination to fly west, while Kosmos-955, like most rockets, was launched east (using the Earth's rotation as additional thrust).

Peculiar perforations

Some Petrozavodsk locals reported the appearance of coin-sized holes in windows during the incident.

It is possible, however, that these holes were already present and noticed only when residents intently peered through the glass to view the event.

Curiosity killed the careers

The Soviet Institute of Sciences tasked experts with addressing the quandary. After a unenthusiastic investigation, their findings were unenlightening. One researcher, Vladimir Migulin, later confessed the reluctance stemmed from a fear of tainting one's reputation with 'speculative science'.

CROP CIRCLES

'Crop circles' are when corn fields are mysteriously flattened into circular patterns. They were a topic of interest by the start of the 80s, the era when films such as *E.T.* (1982) re-ignited interest in aliens. A handful of sightings of UFOs committing acts of crop vandalism solidified the phenomenon.

TASMANIA '75

ZÜRICH '75

IRELAND '76

TWYELL '78

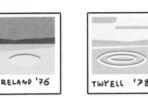

Serious cereology

Investigators known as 'cereologists' specialised in the study of the phenomenon. At the peak of the craze, rural farms that became host to new crop circles could expect the prompt arrival of a frantic parties of cereologists, journalists, sightseers and New Age folk, all clamouring for access to the cryptic crops.

Work of the Devil?

At the time they were first reported in 17th century England, many believed crop circles to be created by devils or færies. This was also the explanation for 'elf' or 'færie rings', the circular fungal growths that may be seen in grassy areas to this day.

Artists or aliens?

In 1991, artists Doug Bower and Dave Chorley announced that were responsible for many of the crop circles around England since the late 1970s. In front of an eager audience of press and cereologists, they demonstrated how rope and wooden planks were used to make crop circles. There of course remains a stubborn camp of those who believe it was UFOs.

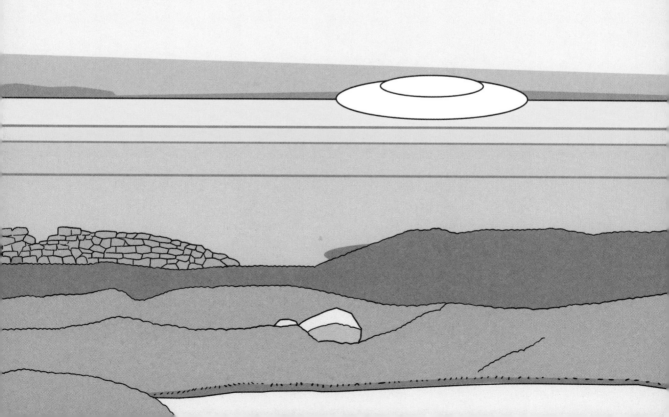

THE 1980S

Munich, Germany, July 1984

CAUGHT ON CAMERA

The 1980s were perhaps the last gasp of widely circulated UFO photography. As the public became suspicious of easily achieved visual effects, press reportage seems to have declined. From the beginning, many photos have been identified as forgeries. Like ghost photography, tricks may be used to create convincing or comically poor paranormal documentation.

Los Angeles, USA, 1942
This UFO was spotted by American anti-aircraft batteries and fired upon over the course of an hour. It was probably a stray balloon.

McMinville, Oregon, USA, 1950
Long considered genuine by many ufologists, skeptics believe the photo features a model hung by string.

Lossiemouth, UK, 1954
Likely copycat model of the 'Adamski-type' flying saucer.

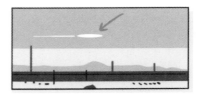

Holloman Airforce Base, USA, 1957
A silver object. Project Blue Book judged the object to be a blurred image of an aeroplane.

Falcon Lake, Canada, 1967
Photo of Stefan Michalak's burn marks, which he claimed were inflicted by a flying saucer.

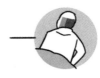

Burgh Marsh, UK, 1964
Apparently a space-suited alien, or more likely an effect of over-exposure.

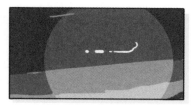

Hessdalen, Norway, 1982,
*A picture of the repeatedly
documented local phenomena, judged
to be some weird atmospheric event.*

Gulf Breeze, USA, 1987
*Almost certainly a model flying saucer,
which was recently found in the attic
of the photographer.*

Ilkley, UK, 1987
*A short alien on the Yorkshire moors,
perhaps looking for a tearoom? Skeptics
say the object is a mannequin.*

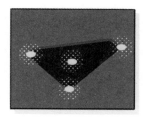

Wallonia, Belgium, 1990
*A triangular spaceship, later revealed
by the photographer to be made of
polystyrene and light-bulbs.*

Phoenix, Arizona, USA, 1997
*A large triangular UFO, sceptics
believe it to be an USAF flare exercise.*

Antarctic, reported in 2018
*A crashed flying saucer found on
an internet map browser, or perhaps
a chunk of rock that slipped down
the mountain range.*

La Junta, USA, 2019
*Home security footage that seems to
depict an alien or some type of færie.
It is most likely a child with underwear
on their head.*

North Carolina Coast, USA, 2019
*Footage of lights taken by a ferry
passenger. Most likely a Navy
flare exercise.*

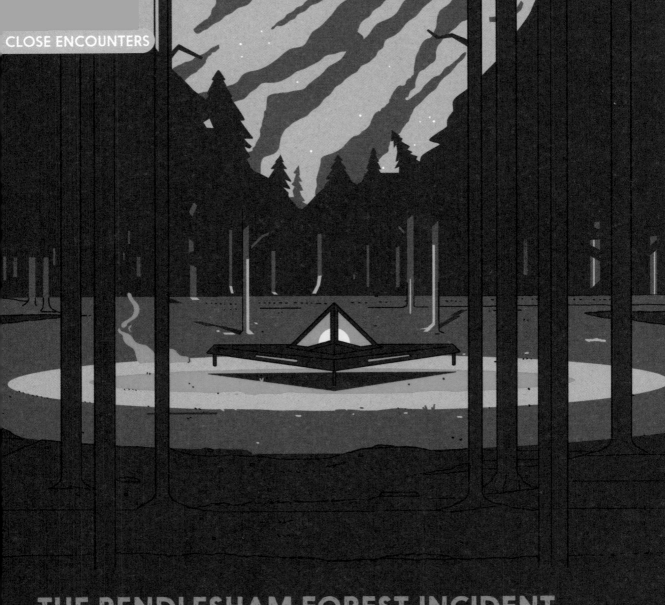

THE RENDLESHAM FOREST INCIDENT

During Christmas 1980, the US Air Force base in Rendlesham, UK, was host to an unexpected festive guest. In the very early hours of Boxing Day, patrolling soldiers were astonished to discover a triangular craft parked in the woods.

The witness accounts vary, but one soldier felt the craft broadcast numbers directly into his mind, before whizzing off back into the sky. Returning later with police backup, the soldiers found broken branches and apparent landing marks on the ground

Extraterrestrial encore

Extraordinarily, the UFO came back that night. Elated soldiers interrupted Boxing Day dinner festivities to announce they had seen another light in the woods. A few soldiers hastily gathered an arsenal of recording equipment and went in pursuit.

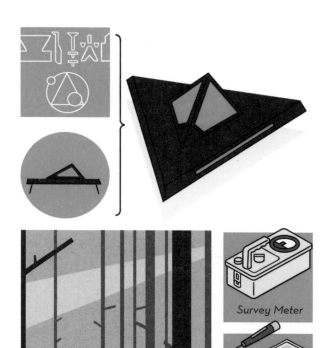

While the soldiers have since given slightly differing accounts, they managed to observe unusual broken foliage and apparent heat traces of the object. Finally, in the early hours, they pursued a light above the tree-line. Remarkably, this chase was recorded on audio tape, 20 minutes of which is freely available online, and is quite a thrilling listen. Listeners may hear the expletive-ridden commentary of the soldiers as they observed the UFO's movements and bright colours before it disappeared.

Survey Meter

Tape recorder

Just a lighthouse?

Prominent skeptic Ian Ridpath suggested personnel excitedly misinterpreted a range of nocturnal lights (a meteor, bright stars, and the nearby lighthouse) as a single event. Key witness John Burroughs has since admitted he had never seen the lighthouse before, suggesting the US personnel were unfamiliar with the landscape beyond their base.

WHITLEY STRIEBER'S COMMUNION

In the fresh hours of 26th December 1985, American horror writer Whitley Strieber woke to the sight of a creature lurking by his bed. The next day he recalled terrifying, disjointed visions of an abduction.

With help from hypnotist Dr Klein, Strieber explored his memories of abduction and wrote the bestselling book *Communion* (1987), a memoir detailing his memories of being subject to painful medical tests at the hands of alien-like creatures.

Art Bell, radio host

Highs and lows

Strieber appeared frequently on talkshow *Coast to Coast AM*. The host Art Bell was beloved for providing a platform for unusual guests and topics (commonly ufologists and UFOs, respectively). The animated show *South Park*, however crudely mocked the medical 'rectal probing' Strieber recalled being victim to. This problematic gag has became shorthand for mocking abductees.

HESSDALEN LIGHTS

Since the 1930s, Hessdalen in Norway has been host to reoccurring lights flitting around the valleys. The lights appear day and night, move at incredible speeds and perform dazzling manoeuvres.

Formal investigation of the lights began in the 1980s, with the efforts of civilian organisations UFO-Norge and UFO-Sweden. The research continues to this day, and is one of the most consistently recorded UFO phenomena. A permanent and automated observatory called 'Hessdalen AMS' continues to monitor the unusual light activity. Present theories suggest the lights are the product of ionised gasses, or piezoelectricity (rocks ejecting balls of energy due to unique chemical properties).

Researchers have measured the phenomena with a vast array of equipment:

Seismometer

Geiger counter

Spectrum analyser

IR viewer

RADAR

Magnetograph

IN DEEP WATER

A tangent of ufology involves the study of 'Unidentified Submerged Objects' (USOs). These USOs have been documented for as long as there have been ships on the sea and range from sea monsters to flying saucers that launch out of the ocean.

The Russian red sphere

In 1965 the crew of the Soviet steamship *Raduga* was stupefied to see a bright red sphere emerge from the sea some two miles from their ship. The sphere hovered above the waves, illuminating them with crimson light for several minutes before returning to the depths. Similar cases have been reported by ships throughout time. Some appear to be balls of light, while others have been described as solid silvery craft.

Loch Ness monster

The Loch Ness monster is a cryptid said to haunt the titular Loch Ness in Scotland. First appearing in the 19th century, it was later popularised in the 1930s by the famous 'Surgeon's photograph' which is now recognised as a forgery. Despite this, many have sworn to have encountered 'Nessie' and Loch Ness remains a popular tourist destination. Since the 1980s some ufologists have suggested that the Loch Ness monster is an abandoned alien or discarded hybrid experiment.

Shag Harbour incident

In 1967 at least eleven witnesses in rural Canada saw a UFO fall into the waters of Shag Harbour. Within 15 minutes of the crash Canadian police rowed out in an attempt to recover a floating object, only to watch it sink into frothing yellow foam. Military divers later attempted to recover the object but their efforts were unfruitful. Like Loch Ness, Shag Harbour remains a paranormal tourist location.

Fast Mover

In 2017, ufologist Marc D'Antonio claimed to have been treated to a ride in a US Navy submarine, during which sonar operators got excited by a 'Fast Mover'. This is apparently the term for an unknown, underwater object travelling at high speed. At the time, the story was given significant attention in tabloid newspapers, but little confirmation has followed.

AREA 51

Commonly known as Area 51, the USAF Groom Lake Facility in Nevada, USA is often associated with UFO conspiracies. In 1955, the base was established as a test facility for stealth aircraft and captured Soviet planes. Sceptics have suggested these tests may account for historic UFO sightings in the area.

Weird whistle-blowers

In the 1980s individuals came forward claiming insider knowledge of government or military dealings with aliens. These informants belong to the same category as contactees and abductees, relying on the strength of their lucid disposition to convey supposed truths. On 14th May 1989, Bob Lazar appeared as an informant on Las Vegas TV. Lazar disclosed his work in back-engineering a flying saucer at the secret facility 'S4' near Area 51. According to Lazar, he was recruited after receiving prestigious degrees from MIT and Caltech.

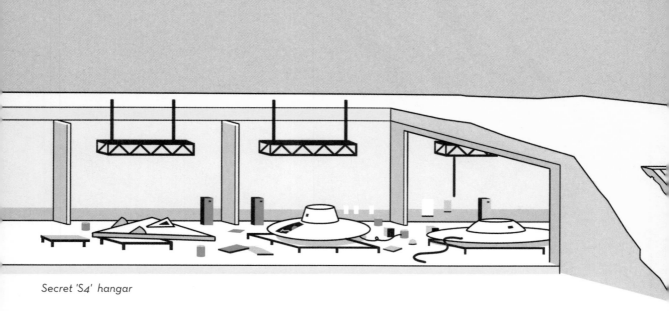

Secret 'S4' hangar

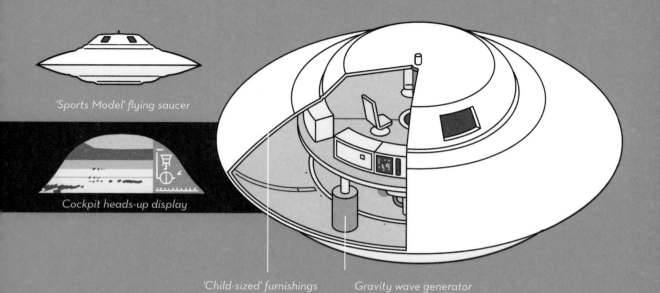

'Sports Model' flying saucer

Cockpit heads-up display

'Child-sized' furnishings

Gravity wave generator

Sports Model

Lazar claimed his employment at S4 involved studying what he called the 'sports model' flying saucer (so named due to its sleek design). He also recalled briefing documents that made reference to the original owners of the craft: grey aliens from Zeta Reticuli.

Friend or Fraud?

Ufologists such as Stanton Friedman called into question Lazar's academic background, for which they found no evidence. This led them to believe his tales of Area 51 were false. Lazar was one of many esoteric informants of the era. There were even alleged time travellers like John Titor, who posted on the early internet, making claims about the future.

THE 1990S

Lake Backsjön, Sweden, July 1999

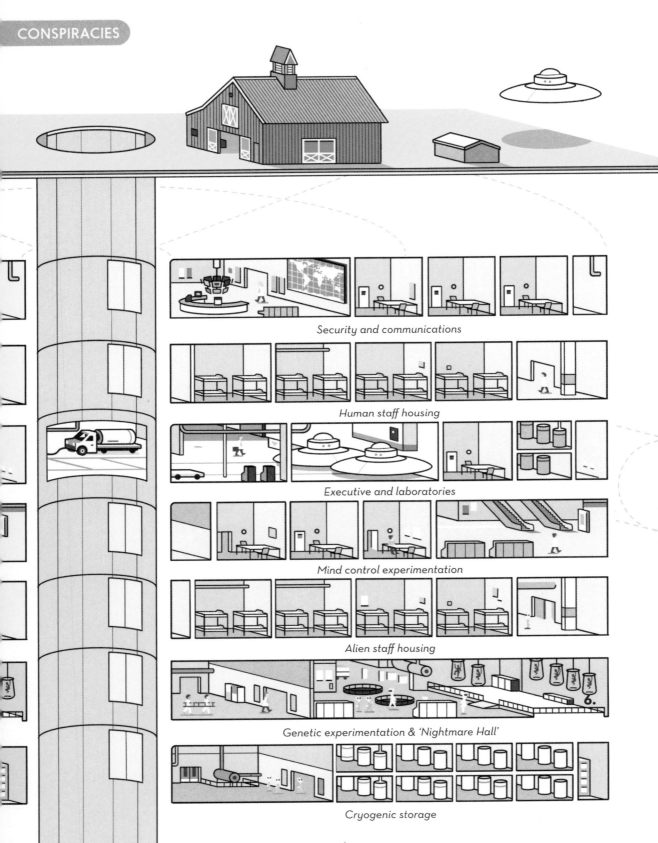

Security and communications

Human staff housing

Executive and laboratories

Mind control experimentation

Alien staff housing

Genetic experimentation & 'Nightmare Hall'

Cryogenic storage

DULCE BASE

In lieu of major new UFO events, the 1990s were full of fringe conspiracy theories facilitated by UFO conventions and early internet message boards. Dulce Base is an alleged secret underground facility in New Mexico, USA, used by the US government and aliens to engage with unspeakably horrid mind control and genetic experiments.

Grey

Mystery meat mixture described by Schneider (it's people...)

Secret signals
Conspiracy theorist Paul Bennewitz helped launched the Dulce legend in the late 1980s, after he apparently received radio transmissions from the base. Many in the UFO community believed him to be suffering from paranoid delusions.

Insider information
In 1995, Phillip Schneider claimed to have worked at Dulce Base in the 1970s. He discussed working with grey aliens, reptilians and human staff until a battle broke out over the treatment of human test subjects. This 'Battle of 1976' apparently resulted in the deaths of over 60 people at the hands of grey aliens.

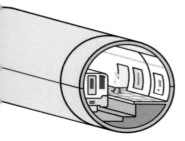

Secret train to nearby town of Los Alamos

Dulce's department logos

'Trilateral'   Dulce Base   US Govt.

Prepare for unforeseen consequences
The magnificent imagery of Dulce Base, as described by Schneider, has been highly influential on science fiction. Underground bases full of alien technology and creepy experiments have become a trope in many video games such as *Half Life* (1998).

THE ALIEN AGENDA

In the 1990s, alien conspiracies were spread by fledgling internet message boards and books by high profile pundits. These narratives tied together details of abduction cases and supposed telepathic contact with aliens.

US president Eisenhower apparently met aliens in 1955

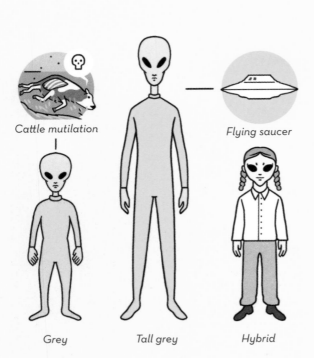

Cattle mutilation

Flying saucer

Grey

Tall grey

Hybrid

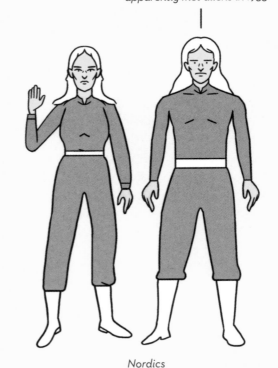

Nordics

Greys
Short greys from Zeta Reticuli are the blue-collar workers of the cosmic conspiracy. They carry out abductions and medical experiments on humans and animals.

Tall greys
Abductees have witnessed these lanky aliens commanding their shorter staff members. They probably earn more too.

Grey-human hybrids
Theorists believe the endgame of grey abductions is the production of alien-human hybrids, which are slowly being introduced into Earth's population, for reasons best known to them.

Nordics/Pleiadians
Hailing from the Pleiades star cluster, these Scandinavian-looking aliens are thought to be mostly benevolent. Nordics intersect with conspiracy theories, having apparently signed a treaty with US president Eisenhower in 1955. Nordics were most commonly witnessed in the 1950s, and depending on one's perspective were inspired by, or were the inspiration for, aliens in films like *The Day the Earth Stood Still* (1951).

Hollow Earth residence

Anyone could be a lizard!

Fashionable hood outfit

Reptilians

Reptilians

Malevolent reptilian aliens are said to be behind pretty much all worldly misery as they apparently harvest mortal pain as a renewable energy resource. Furthermore, shape-shifting reptilians have supposedly infiltrated all spheres of influence as celebrities and politicians. It is said that Reptilians also lurk in the centre of the Earth so they can plot against us in private.

Nordic: *Lawful good*

Avians: *Neutral good*

Alien gods: *Chaotic good*

Mantises *Lawful neutral*

Humans: *Neutral*

Hybrids: *Chaotic neutral*

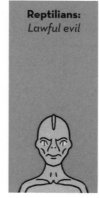

Reptilians: *Lawful evil*

Men in Black: *Neutral evil*

Greys: *Chaotic evil*

The battle for Earth

The more commonly accepted theories state that *at least* three alien species are vying for dominion over Earth; the **Greys**, **Nordics** and **Reptilians**. Other theories throw in mantises, avians and space dragons for extra colour.

UFOS ON TV

As with film, television has long reflected the public's interest in UFOs.

The invasion will be televised

With few exceptions, flying saucers were depicted on TV as simplistic vehicles of malevolent aliens. Broader interest in UFOs declined until the 1990s when *The X-Files* introduced obscure UFOs and broader paranormal theories to a mass audience.

The Twilight Zone (1959)
Early appearances of flying saucers.

Doctor Who (1963)
The main antagonist aliens, called Daleks, get about in flying saucers.

The Outer Limits (1963)
Featured grey aliens.

The Invaders (1967)
Prominently featured flying saucers.

STV broadcast interruption (1977)
An alleged alien took over the audio of a UK news broadcast to warn humanity about the folly of war.

V (1983)
Featured aliens and flying saucers with an aspect of conspiracy theory.

Twin Peaks (1990)
A blend of occult, Men in Black, and UFO lore. Project Blue Book is featured in the narrative.

Star Trek: Deep Space Nine (1993)
Quark, an alien barkeep, accidentally becomes the cause of the Roswell incident.

The X-Files (1993)
Showrunner Chris Carter devised this series to explore UFO and paranormal history. Protagonist FBI agents Fox Mulder and Dana Scully (the latter named after ufologist Frank Scully) investigate crimes with a spooky flavour. Cryptids, psychic powers, UFOs and ghosts are all featured, along with pretty much every conspiracy theory folded into an over-arching alien subplot. When the X-Files began airing, many UFO organisations found their memberships swelled dramatically.

Star Gate: SG1 (1994)
Featured an Area 51-like base and various aliens.

THE ARIEL SCHOOL INCIDENT

On 16th September 1994, the children of Ariel School in Ruwa, Zimbabwe, witnessed the extraordinary landing of a spaceship. They described seeing two alien figures briefly emerge, and after some consideration, return to their ship and speed off.

Matchstick men

After the visitation, teachers encouraged the children to draw what they had seen. The drawings were remarkably similar in content and detail.

The school became the object of mass media and ufologist attention, leading skeptics to speculate that the stress of attention had caused the children to create false memories.

ALIEN AUTOPSY

In May 1995, filmmaker Ray Santilli presented a film that allegedly featured the alien cadavers recovered from the Roswell incident.

The film was central to the highly viewed TV special *Alien Autopsy: Fact or Fiction*. After the broadcast, experts weighed in and declared the film fake, noting that the surgeons were holding equipment incorrectly. Special effects artists also recognised the alien flesh as simple rubber.

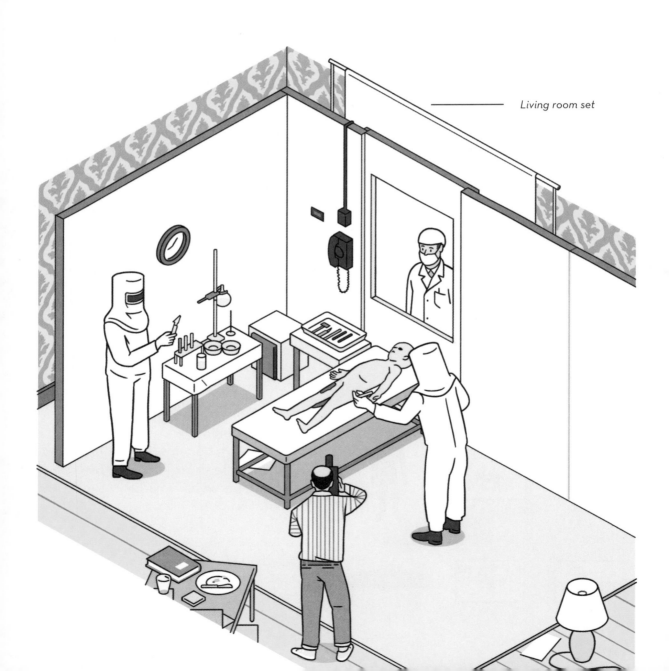

Living room set

Hoax!

In more recent years, crew members of the production have disclosed the film was a hoax made in a London flat. Apparently, Santilli sourced period-accurate equipment for the film, including the case the footage was stored in. The alien was based on the cast of child, with various meats bought from a local butchers to produce the gore.

Regardless, the footage made a significant cultural impact and has been much-referred to and pastiched in the media. The story behind the hoax was made into the feature film *Alien Autopsy* (2006). When promoting the film, Santilli suggested the original footage was a 'restoration' based on real footage he acquired.

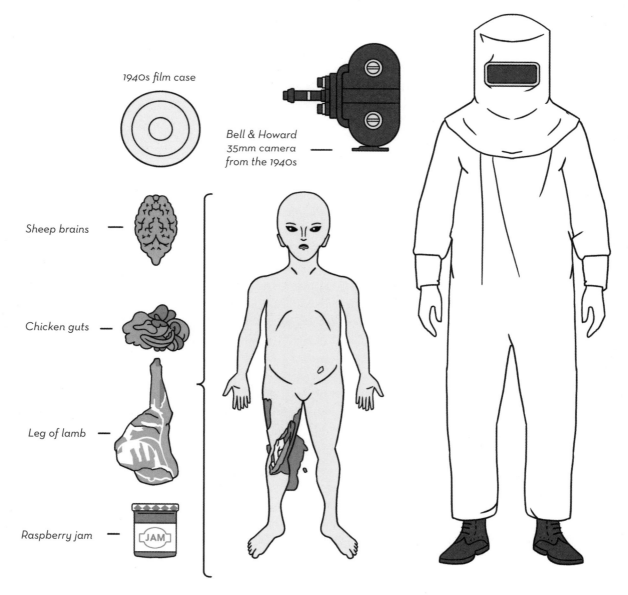

1940s film case

Bell & Howard 35mm camera from the 1940s

Sheep brains

Chicken guts

Leg of lamb

Raspberry jam

JAM

THE PHOENIX LIGHTS

On the night of 13th March 1997, residents around the state of Arizona witnessed two unusual events. The first event – shortly after 8pm – was described as a gigantic V-shaped craft with numerous lights drifting directly over Phoenix, Arizona, USA. The second event, around 10pm, involved nine lights that seemed to hover over Phoenix. This event was well-documented on video by onlookers.

Many witnesses were frustrated and unimpressed by the tone of reportage and the official explanations, and formed their own investigations and support groups. The Air Force claimed responsibility for the events, stating the lights had been flares dropped from an aircraft during a training exercise. Arizona's governor at the time, Fife Symington III, saw the funny side and held a press conference in which he poured scorn on a person in an alien costume.

The Phoenix lights are perhaps the best-documented contemporary mass-UFO sighting, a reminder that UFO witnesses are usually ordinary folk.

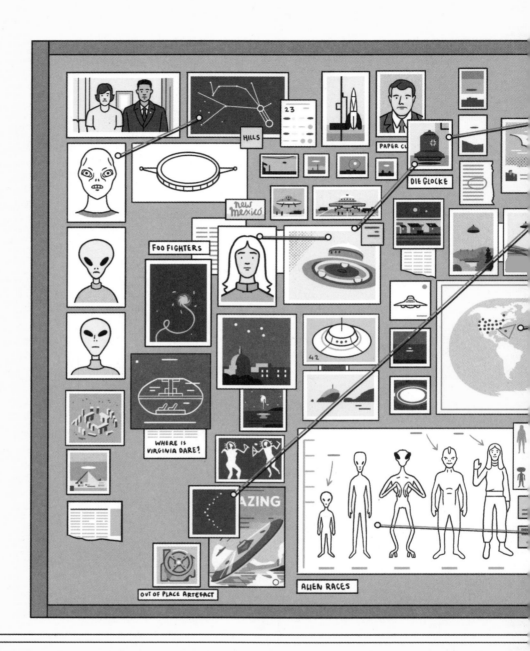

GOVERNMENT DISCLOSURE

Today, many ufologists believe their most productive activity is campaigning for the governmental disclosure of official UFO information. It is hoped that transparency will help reduce ridicule of the subject and encourage more experienced scientists to apply themselves to the field.

Critics of the disclosure movement suggest that in lieu of frequent and new UFO incidents, career ufologists have turned to paranoid conspiracies about the government to argue about instead.

A large UFO enthusiast conference

Small public event

Podcast

Internet pundit

Hack the planet

In 2002, Scottish UFO enthusiast Gary McKinnon was arrested on suspicion of hacking into US military computers and causing unlawful damage. McKinnon said that his motive was to find secret UFO evidence. After lengthy proceedings, extradition to the United States was blocked by the UK government and he has since been free to give interviews regarding the secret files he claims to have been privy to, namely pictures of UFOs and suspicious spreadsheets.

Memes and clickbait

Online, the disclosure movement and broader UFO culture is often the subject of low quality clickbait and ironic humour.

The Area 51 raid

On 20th September 2019, a large group of good-humoured folk arrived at Area 51 to, quote, "see them aliens". The raiders took pictures and enjoyed an impromptu music festival. Unfortunately, the raiders were unable to secure the release of any alien prisoners.

THE PENTAGON'S SAUCER SEARCH

The US's intelligence headquarters, known as the Pentagon, began the Advanced Aerospace Threat Identification Program (AATIP) in 2007. The purpose of this project was to study 'Unidentified Aerial Phenomena' (UAP), a contemporary label for UFOs.

The program was revealed in 2017, when footage of UFOs tracked by Navy jets was leaked. Officials provided conflicting statements about the purpose of AATIP, until they finally confirmed the validity of the videos in 2020.

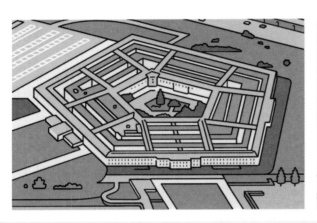

Luis Elizondo,
AATIP Director 2007–2012

The *USS Nimitz* incident

In November 2004, the US Navy tracked a UFO off the coast of California. Two Super Hornet jets from the *USS Nimitz* were sent to intercept. The pilots were startled to discover a white oblong-shaped UFO hovering above churning waves. Thankfully, they were able to record their pursuit of the object on their infrared cameras; this was leaked online.

Excessive acronyms

The US intelligence community is a confusing web of agencies and external contractors. AATIP awarded a research contract to Bigelow Aerospace Advanced Space Studies (BAASS). The owner, Robert Bigelow, impressed Pentagon folk with talk of his allegedly haunted estate, known as 'Skinwalker Ranch'.

Clandestine continuation

While it was reported that AATIP closed down in 2012, insiders say that it is still running in some form. The Pentagon remains typically evasive on the matter when pressed for answers by journalists.

BAASS's review of UAPs is a riveting read

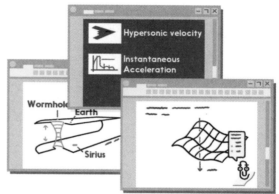

SETI

The Search for Extraterrestrial Intelligence (SETI) is the collective effort of many scientists monitoring different types of transmissions from space and making note of any that seem more coherent than random background noise. If we find irrefutable evidence of extraterrestrial life, it is supposed by many that it will be revealed by a radio telescope rather than a flying saucer landing on the lawn of a surprised world leader.

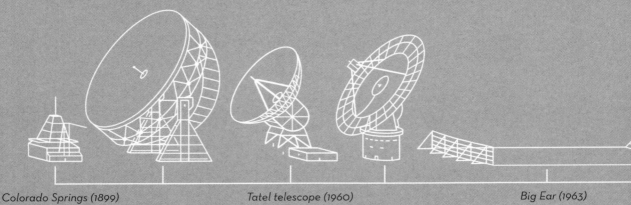

Colorado Springs (1899)

Jodrell Bank (1957)

Tatel telescope (1960)

Parkes Observatory (1961)

Big Ear (1963)

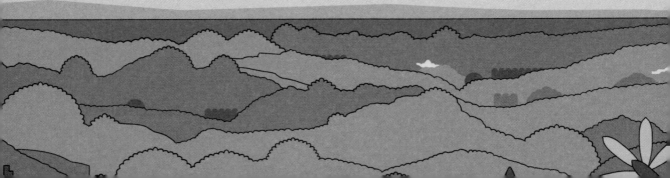

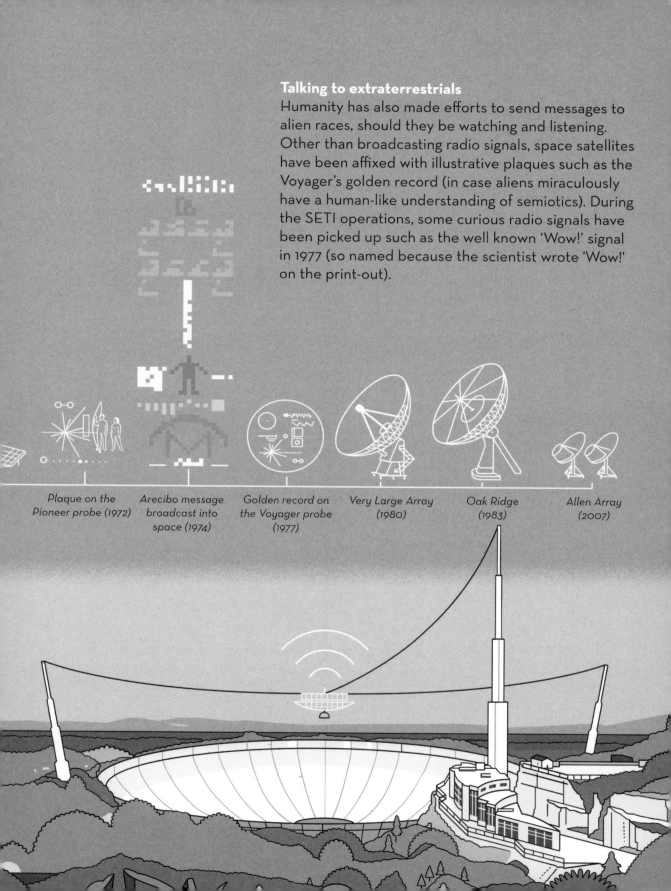

Talking to extraterrestrials

Humanity has also made efforts to send messages to alien races, should they be watching and listening. Other than broadcasting radio signals, space satellites have been affixed with illustrative plaques such as the Voyager's golden record (in case aliens miraculously have a human-like understanding of semiotics). During the SETI operations, some curious radio signals have been picked up such as the well known 'Wow!' signal in 1977 (so named because the scientist wrote 'Wow!' on the print-out).

Plaque on the Pioneer probe (1972)

Arecibo message broadcast into space (1974)

Golden record on the Voyager probe (1977)

Very Large Array (1980)

Oak Ridge (1983)

Allen Array (2007)

UFOLOGY TODAY

The interest in UFOs and ufology endures thanks to the contributions of investigators, books, conferences, conventions, forums, podcasts, popular culture and a mass of online pundits.

The study of UFOs is a practice full of passionate and colourful individuals. Certainly, it takes a degree of courage and intensity to investigate a topic so thick with unnavigable swamps of lore and esoterica.

All is not rosy, however. Like other social communities, ufology has had its share of figures unveiled to be sexist, racist or otherwise dreadful. Skeptics continue to malign ufology as a pseudo-science, throwing it in with ghost hunting or extrasensory perception (ESP) research. They perhaps unfairly undermine the practice based on the actions of more passionate factions.

Vocal critics believe the study of UFOs peaked many decades ago with governmental investigations like Project Blue Book, and suggest that there has been a slow decline in research and an increase in introspective conspiracy.

Regardless, there are large amounts of new UFO cases every year. Recent disclosures by officials are a beacon of hope for those that desire government transparency on the matter, helping to de-stigmatise the field of study.

To avoid the most tiresome conclusion to the UFO narrative, 'we may never know', I shall instead issue the following advice: watch the skies and for goodness' sake bring a decent camera.

Individuals associated with UFOs

1. Raymond Palmer (editor)
2. Kenneth Arnold (witness)
3. William 'Mac' Brazel (witness)
4. Nathan Twinning (Blue Book)
5. J. Allen Hynek (Blue Book)
6. John Spencer (ufologist)
7. George Adamski (contactee)
8. Gray Barker (ufologist)
9. Edward Ruppelt (USAF)
10. George Van Tassel (contactee)
11. Donald Keyhoe (ufologist)
12. John G. Fuller (ufologist)
13. Betty & Barney Hill (abductees)
14. Thornton Page (astronomer)
15. Millen Cooke (author)
16. Coral & Jim Lorenzen (APRO)
17. Budd Hopkins (author)
18. Lionel Beer (BUFORA)
19. Chris Styles (ufologist)
20. Don Ledger (ufologist)
21. Frank Scully (ufologist)
22. Travis Walton (abductee)
23. Robert Sheaffer (sceptic)
24. Bob Lazar (alleged saucer engineer)
25. Jacques Vallée (ufologist)

26. Trevor James Constable (ufologist)
27. David Clarke (journalist)
28. Stanton Friedman (ufologist)
29. Nick Pope (ufologist)
30. Jenny Randles (ufologist)
31. Phil Schneider (ufologist)
32. Ardy Sixkiller Clarke (author)
33. Claud Poher (CNES)
34. Arne Gjärdman (FOA)
35. Chris Rutkowski (author)
36. Jaime Maussen (presenter)
37. Linda Moulton Howe (ufologist)
38. Erich Von Däniken (author)
39. Whitley Strieber (abductee)
40. Phillip Class (skeptic)
41. Leslie Kean (journalist)
42. Michael D. Swords (author)
43. Giorgio A. Tsoukalos (presenter)
44. Georgina Bruni (author)
45. Richard Dolan (ufologist)
46. Farah Yurdozu (ufologist)
47. Steven Greer (ufologist)

GLOSSARY

Abductee
Person who claims to have been taken by UFO occupants for invasive testing or a fun transcendental experience.

Area 51
United States Air Force base. The Lockheed U-2 plane was tested here during the Cold War. Some associate the base with the testing of recovered UFOs and other spooky activities.

Artefact (in photography)
Effect or object on an photograph created by dust or strong light sources.

Cattle mutilation
The phenomenon of cattle and other animals being mutilated and killed in an unusual fashion. Some believe aliens are to blame.

Contactee
Person who claims to have either met or had telepathic communication with extraterrestrial beings.

Crank/crackpot
Pejorative term for someone that has extremely unorthodox views.

Crop circle
Crops flattened into creative patterns, usually attributed to inventive artists and pranksters.

Foo Fighter
Small, bright UFOs, witnessed around planes during the Second World War.

Flap
Large increase in UFO sightings, often catalysed by a well-reported event.

Flying saucer
Description applied to UFOs, commonly used since Kenneth Arnold's sighting in 1947 above Mt. Rainer.

Grey
Alien being, commonly reported in sightings and by abductees.

High strangeness
Term coined by Professor J. Allen Hynek, referring to extremely peculiar details from UFO cases.

Hoax
Deceptive story intended to mislead.

Hynek Scale
Professor J. Allen Hynek's method of classifying UFO witness statements. Statements are sorted into close encounters of the first, second or third kind.

Implant
Anomalous object found on an abductee's person after their alleged experience.

Mass hysteria
Group panic based on fear, producing collective illusions and obsessive behaviour.

Men in Black
Malevolent agents that reportedly stalk, threaten or question folk who have seen a UFO. Commonly initialised as 'MiB'.

Missing time
Experience where abductees have reported gaps in memory, in which hours or even days may pass unaccounted-for.

Prank
Less malicious form of hoax, usually motivated by humour.

RADAR
Acronym for 'RAdio Detection And Ranging'; a system for detecting objects by broadcasting radio waves. The waves that are reflected back are interpreted to determine the size and speed of objects.

Prosaic explanation
Mundane explanation for the extraordinary.

Pseudoscience
Practice that is considered scientific or based in method, but is maligned by a lack of evidence or basis in observable facts.

Sceptic
Individual that values scientific inquiry and methodology.

Sleep paralysis
Common explanation for abduction experiences, in which folk are unable to move and experience fear and hallucinations when falling asleep or waking up.

Swamp gas
Natural gasses produced by rotting vegetation. The gasses occasionally produce flying fireballs, which may explain various UFO sightings.

Spooklight
Anomalous light that reappears in the same location. Explanations offered tend to involve aircraft or car headlights.

UAP
Unidentified Aerial Phenomena – the modern parlance for 'UFO' preferred by US officials.

UFO
Initialism for 'Unidentified Flying Object'. Refers to any unrecognised thing in the sky.

Ufologist
Investigator or student of UFO phenomena.

USO
Initialism for 'Unidentified Submerged Object'. Refers to peculiar and unrecognised objects underwater.

Weather balloon
Explanation for various UFO sightings and crashes; including the Roswell incident in 1947.

    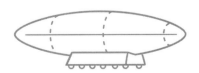

FURTHER READING

 BARKER, Gray, *They Knew too Much About Flying Saucers* (1956), University Books, New York, NY, USA

 BERLITZ, Charles & MOORE, William, *The Roswell Incident* (1980), Crosset and Dunlap, New York, NY, USA

 BLUNDELL, Nigel & BOAR, Roger, *The World's Greatest UFO Mysteries* (1983), Octopus Books Ltd, London, UK

 BYRANT, Helen & Reeve, *Flying Saucer Pilgrimage* (1957), Amherst Press, Amherst, WI, USA

 CLARKE, David, *The UFO Files* (2012), Second edition, Bloomsbury, London, UK

 HYNEK, J. Allen, *The UFO Experience: A Scientific Inquiry* (1972), Abelard & Schuman Ltd., London, UK

 JESSUP, M.K., *The Case for the UFO* (1957), Varo Edition, CreateSpace, Scotts Valley, CA, USA

 KEAN, Leslie, *UFOs: Generals, Pilots, and Government Officials Go on the Record* (2010), Harmony Books, New York, NY, USA

 KEEL, John A., *The Mothman Prophecies* (1975), Saturday Review Press, New York, NY, USA

 KLASS, Phillip J., *UFOs Explained* (1976), Vintage Books, New York, NY, USA

 LESLIE, Desmond & ADAMSKI, George, *Flying Saucers Have Landed* (1953), Werner Laurie, London, UK

 MICHEAL, John & RICKARD, Robert, *Phenomena: A Book of Wonders* (1977), Thames and Hudson, London, UK.

 MICHEALS, Susan, *Sightings* (1996), Fireside, New York, NY, USA

 RANDLES, Jenny, *The UFO Conspiracy* (1987), Javelin Books, London, UK

RUPPELT, Edward J., *The Report of Unidentified Flying Objects* (1956), Doubleday & Company Inc., Garden City, NY, USA

SCULLY, Frank, *Behind the Flying Saucers* (1950), Henry Holt and Company, New York, NY, USA

SLEMAN, Thomas, *Strange But True* (1998), London Bridge, London, UK

SPENCER, John, *The UFO Encyclopedia* (1991), Headline Publishing PLC, London, UK

SWORDS, Michel, *UFOs and Government, A Historical Inquiry* (2012), Anomalist Books, San Antonio, TX, USA

VALLÉE, Jacques, *Forbidden Science* (1992), North Atlantic Books, Berkley, CA, USA

VALLÉE, Jacques, *Messengers of Deception: UFO Contacts and Cults* (1979), And/Or Press, Berkley, CA, USA

VAN TASSEL, George, *I Rode a Flying Saucer* (1952), New Age Publishing, Los Angeles, CA, USA

WILDING-WHITE, Ted, *The World of the Unknown: UFOs* (1977), Usborne Publishing Ltd, Burlington, Canada

WOMACK, Jack, *Flying Saucers are Real!* (2016), Anthology Editions, New York, NY, USA

INDEX

ALIENS THROUGH TIME

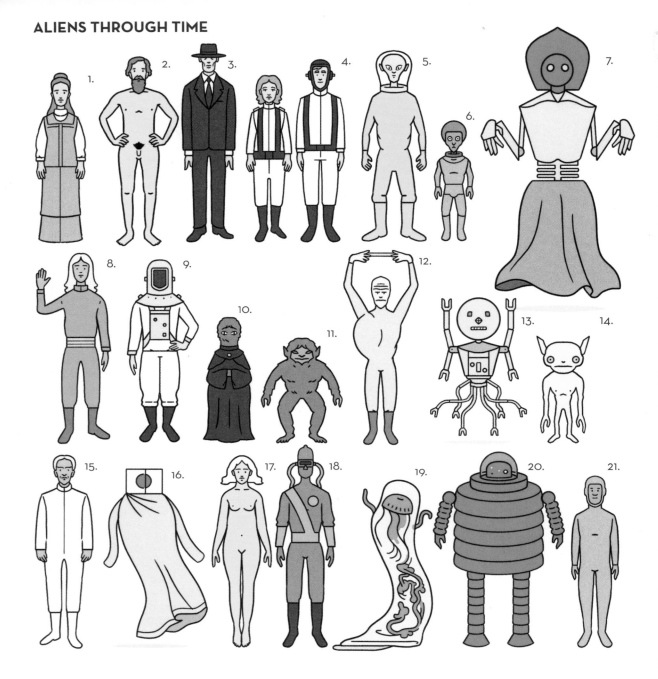